ONCE
WE
WERE

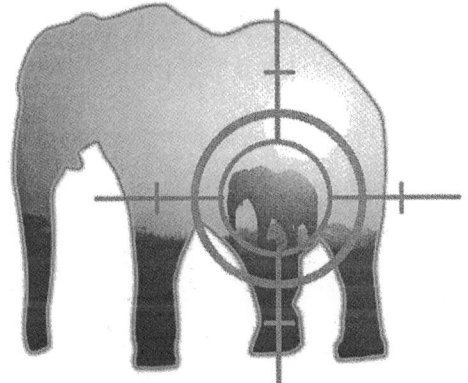

H U N T E R S
A Study of the Evolution of Vascular Disease

ONCE
WE
WERE

HUNTERS
A Study of the Evolution of Vascular Disease

editor : Gianni Belcaro

Imperial College School of Medicine, St Mary's Campus

Imperial College Press

ICP

Published by

Imperial College Press
57 Shelton Street
Covent Garden
London WC2H 9HE

Distributed by

World Scientific Publishing Co. Pte. Ltd.

P O Box 128, Farrer Road, Singapore 912805

USA office: Suite 1B, 1060 Main Street, River Edge, NJ 07661

UK office: 57 Shelton Street, Covent Garden, London WC2H 9HE

British Library Cataloguing-in-Publication Data
A catalogue record for this book is available from the British Library.

ONCE WE WERE HUNTERS
Evolution and Vascular Diseases

ISBN   1-86094-262-8

Printed in Singapore by Uto-Print

PREFACE

This book poses questions such as how well humans are adapted for life today and makes the reader realise how unfit we are for living on our planet today. We are still the hunters we were one million years ago but we no longer hunt. Food is plentiful and all our characteristics that make us good hunters have turned against us. Genes that were essential for our survival as hunters are now our worst enemies, responsible for atherosclerosis leading to heart attacks and strokes. Now we can even manipulate our genetic design and blueprint, and yet the design is still that of a hunter.

The ED/EC (evolutionary designs/ present environmental conditions) disbalance is a major cause of diseases. Many of our primitive instincts confuse us (despite education) into "bad-habits" in terms of modern life. One million years ago the need and the desire could have been essential. Today the need and desire are not essential and may be a danger. Inherited characteristics offering an advantage to the hunter become adverse in the post-reproductive period. Practical advice is offered on altering lifestyle, getting genetic information from the medical family history and on screening programmes. The "cradles of mankind" and our "African Edens", areas which are our evolutionary ground have many lessons for us. It is a joy to read this book. In doing so we understand ourselves better and begin to get glimpses of where we, as hunters, may be heading for.

Andrew Nicolaides
London May 1st, 2001

ACKNOWLEDGEMENTS

Many thanks to:
Nikon
British Airways
Gametrackers
The Royal Society of Medicine
St Mary's Hospital London
Tuttoviaggi, Milan
BBC
The Trans-African Vascular Group
particularly:
    Bob Rutherford
    Wesley Moore
    Andrew Nicolaides
    Frank Veith
    and all our African trips companions and guides. They all gave me ideas and suggestions and their wonderful company in some of the best days (so far) of this life.

CONTENTS

INTRODUCTION

A. The Acid that Eats Anything

Imagine an acid so corrosive that it could attack and destroy anything. Even containers could not keep it. The acid would destroy all of them. Nothing, if this substance is present, would be safe from it.

This is the central idea of Daniel C Dennett's book *Darwin's Dangerous Idea. Evolution and the Meanings of Life*. The dangerous idea is the concept of natural evolution. Once you see and start understanding its basic concepts, nothing is the same any more. From war, peace, the love of Romeo for Juliet, hamburgers and fries, random killers, antibiotics and all the other things and facts, minor or major ones, which are the million bits that make life, they are all seen in a different light. There is a different meaning for each. Everything goes according to evolutionary rules.

As Dennett writes:

> *"Bearing the unmistakable likeness to universal acid, Darwin's idea eats through just about every traditional concept and leaves in its wake a revolutionised world view, with most of the old landmarks still recognisable but transformed in fundamental ways".*

Darwin's idea destroys completely the biblical story of creation and other biblical views, relegating to fascinating mythical stories the idea of a personal God, and attacks even the concept of human mind and consciousness.

> *Was this the face that launched a thousand ships and burnt the topless towers of Ilium?*
> *Sweet Helen make me immortal with a kiss!*

Seen in the light of evolution it is not just the survival of the fittest or supremacy of the most beautiful, it is the attempt to transfer to the next generation genes producing beauty.

Sweet Helen make my genes less mortal with sex.

I do not think that Darwin had, at the time he wrote most of his books, the idea of creating a new philosophy. He stated some facts, elaborated from observations deducted from his contacts and experience with nature. These facts and observations had been there, available to other scientists before him. But he sees all of them in a different light. He manages to link all of them in a single common network and chain of events. The results of selection, the survival of only the most adaptable or the fittest constitute a deep and transversal concept which covers any field of human knowledge from philosophy to biology, from basic empirical science to the most abstract speculation. From age to age, deers run more swiftly, cats stalk their prey more silently and giraffes' necks become longer and longer to get the better leaves. Given enough time, this mechanism could account for the whole long development from protozoa to homo sapiens. However, in this framework, man is now the only being which has outrun evolution. This has happened only in recent times and, during the time of Darwin, the role of humans as a factor which modifies the environment was not yet perceivable.

When planes fly faster than sound, we do not hear the sound they make. As, at one stage, black holes, after increasing and increasing their gravity, produce so much gravity that not even light can go away from them. At one stage, humans became the protagonists of their evolution and started to modify the environment. The modifications are so many and so large that now the most important evolutionary force on the planet is homo. The modification of the environment has been so fast, large, massive and diffuse that humans, over a short period of time, having evolved for millions of years on a planet and under defined life conditions, find themselves in a different planet and in totally different life conditions.

B. De-Evolution

This evolutionary one-animal, selfish process, which may be defined as de-evolution, now affects our planet so much that many species, even those marginally competitive with humans, either adapt themselves to live in a homanised (I prefer this word to humanised) world or disappear and perish. They particularly suffer and disappear if they are specialised to live (after million of years of evolution) in an environment which is even marginally inconvenient to us or to our life conditions. So, foxes and wolves now survive not because they are aggressive but because they are adaptable and learn to eat rubbish. Other animals are killed and became extinct just because the environment has been modified forever by agriculture or by other activities of humans.

C. Baby, We Were Born to Run

After walking and running on a planet for millions of years, 12 hours a day all year round, just to get a little bit of food to survive, we now live in a sedentary environment with a lot of food around us. We were born and evolved in a world with very limited quantities of sugar, fats and proteins and salt. We acquired such a great desire for them because in those pristine conditions, the possibility of recognising and getting them could be the difference between life and death.

Seeing the world with medical glasses, after reading Dennett's book and an article about it in the Independent (*Ray Monk: Acid from the Tree of Life*), I become more and more interested in answering a few questions regarding some of the most common diseases in our post-industrial, de-evolutionised society. If cardiovascular disease is so widespead, is this because there is a discrepancy between our evolutionary design and present life conditions? Can we see diseases in a different light if we consider some of them in the light of evolution? My trips to Africa, some areas of which are possibly still very similar to the environment which was the stage of our pristine life and the evolutionary ground for 99% of all the evolutionary life of humans, had already produced a series of concepts and ideas. We were designed and built to walk 12 hours a day, leading a nomadic life, gathering and hunting, running away from other predators, always on the edge of starvation. In these conditions, the capability of accumulating fats and proteins in a very effective way could have been an important advantage in situations of starvation.

The discrepancy between evolutionary design and present life conditions could be the basis for the most frequent disease of our age, atherosclerosis.

However, other diseases and conditions, such as cancers of the female reproductive system, manic depressive illness and unipolar depression, may have derived their increased frequency in our society as a consequence of the discrepancy between design and the present conditions.

In the post-industrial world, women experience earlier menarche, later first birth (the time between menarche and first pregnancy is increased 4–5 times), less nursing, lower parity and later menopause. The net effect is a more prolonged exposure to oestrogenic hormones which in turn increases cell reproduction. And cells that are dividing frequently are more likely to develop malignancy.

Manic depressive illness (MDI) affects some 1% of all adults substantially lowering fitness (suicide rates are high, some 20%) and causing premature deaths. This frequency suggests that there must be some evolutionary advantage in preserving such a high percentage of individuals with MDI to counterbalance the powerful disadvantages. It is possible that people with high creativity have a high frequency of MDI.

Unipolar depression (UP) has a lifetime rate greater than 10% in post-industrial societies. It is difficult to see which favourable character may be associated with UP. To oversimplify, chronic sadness associated with UP may be seen as a stimulus leading to long-term changes in behaviour as much as pain is associated with short-term changes. It is possible that the adaptive response in a changed environment can overshoot, resulting in a non-adaptive behaviour.

Atherosclerosis is an *accumulation disorder* consisting of a complex set of metabolic alterations often associated with similar accumulation disorders (diabetes, hypertension, obesity). The body, built in and for a situation of *low-calorie-low-sugar-low-salt-low-fat/protein intake* and *not-too-many-contacts* is now in a world of plenty. The positive, adaptive mechanism of accumulating *as-much-as-you-can* is clearly a disadvantage leading to arterial changes. Also, too many contacts will add too much negative stress which may constitute by itself an important factor in promoting cardiovascular disease.

Atherosclerosis is also possibly asssociated to events in foetal life. The foetus makes responses to poor nutrition which, although adaptive in preserving brain growth, can lead to coronary disease in adult life.

D. Once We Were Hunters

So, here we are. No sophisticated experimental design, just speculations. The evidence is not what you would consider and accept in scientific circles as documented evidence.

Still, it may save your life to consider some points. We were born to run, once we were hunters, we still are and to be healthy, to feel *in the right place*, we should think in a different way.

We have to stop changing our environment, our planet, so quickly, so dramatically and in such a meaningless way. Without our evolutionary playground, we would be lost either physically or mentally.

Our evolutionary mates need to have their place around us and we need to understand that we are the combined results of the millions of years of interaction with them; sometimes eating some of them, and sometimes being eaten by some of them.

If you spend some of your time with a mountain gorilla, that is, if you can find one, you may look at him with human superiority for a while before understanding, maybe from his eyes, that he has the same right to have a chance to survive in his place. We need so many things to live and they need so little. We spoil so much of the environment around us in order to have so many useless things. Are they on the top of the evolutionary ladder, so perfect, so good and powerful, so experienced to live with almost nothing?

We need so much to survive and we suppose that we are at the top!

E. Save Your Life

Understanding the meaning of our deep hunting nature may give us a different view of our most common disease, atherosclerosis, and its dis-evolutionary meaning. It could be a great help to fight a clinical problem which affects so many at such a large cost.

The immensity of the problem of preventing and treating an advanced disease could be counterbalanced by the understanding of our deep, real nature and may indicate to us a way of life more compatible with our evolutionary needs and meanings.

> *Hermia: Be it so, Lysander: find you out a bed: for I upon this bank will rest my head.*

Fig. 1 The Lwanga river in Zambia. This place still retains the original characters of our evolutionary playground.

Fig. 2 The lilac breasted roller, one of the most typical birds found in Africa in the remaining edens which were once our evolutionary grounds.

PART I

EVOLUTION AND CARDIOVASCULAR DISEASES IN A NUTSHELL

Chapter 1

LUCY IN THE SKY

Location: Awash, Ethiopia

Year: 1974

Main Characters: Prof. Johanson & co-workers, Lucy (dead, very dead, bones only).

Musical Background: Lucy in the sky

<pre>
 F   F7   Bb   Bbm   F   F7
</pre>
Picture yourself in a boat on a river with tangerine trees and

<pre>
Bb   Db
</pre>
marmalade skies.

<pre>
 F   F7   Bb   Bbm   F   F7
</pre>
Somebody calls you answer quite slowly a girl with a kaleidoscope

<pre>
Dm   D7
</pre>
eyes

<pre>
 G   A9   D7   G
</pre>
Cellophane flowers of yellow and green, towering over your head.

<pre>
A9   E   Bm
</pre>
Look for the girl with the sun in her eyes and she's gone.

<pre>
E   A   B
</pre>
Lucy in the sky with diamonds
Lucy in the sky with diamonds
Lucy in the sky with diamonds Ah (C); ah (last time, B)

Late Summer 1984 August 23rd

Kamoya Kimeu spotted a small fragment of an ancient cranium lying among some pebbles on a slope near a dry ravine worn by water from a seasonal stream. After five seasons of excavations following this finding, tons of sediment were removed and the complete skeleton of an individual who had died at the edge of an ancient lake more than 1.5 million years ago was recovered. The Turkana boy was about nine years old when he died and the cause of his death is still a mystery. The skeleton was something very special. Nothing as complete as this million-year-old skeleton had been found before in human fossil records, apart from Neanderthal times (which is much more recent, only some 100,000 years before our time).

The skeleton of the boy and his (speculated) missing fleshy parts indicate that years of evolution, in relatively ancient times, had already produced an individual remarkably similar to us. His body was perfectly designed for a hunting and gathering life in an environment mainly constituted by warm tropical or sub-tropical savanna, already present at the time the individual lived in that area. The result of millions of years of evolution and the emergence of the design of this ape-derived human are clearly visible. After this finding, another long period of one and half million years had been used by evolution to perfect that excellent design in very similar life conditions.

And Now?

When the environment had finally produced the most perfect hunter design (perfect for hunting in these conditions), the hunter started to change his environment, altering or abolishing forever most of the powerful evolutionary forces at the basis of his structural design.

We are now still very similar to that boy and to our ancestors living one million years ago but we live (at least most of us in the post-industrial society) in a totally different environment which has quickly grown around us.

We have strongly modified in less than 100 years, and even more and faster in the last 25 years, the stage where we act. If our evolution goes back to some five million years (50,000 years is 1% of five million), this last period of 100 years is equivalent to 0.002% of our evolutionary period, and this is nothing.

We Still are Hunters

However, our stage and living territory is not a hunting ground any more. But we are deeply rooted to our evolutionary past and our life is affected by our evolutionary history. We are now out of place or maybe the place is not good for us any more.

Fig. 3 An Australian hunter in Darwin's book *Journal of a Voyage Around the World.* Most natives were killed by white men or diseases in a short period after the discovery of Australia.

Chapter 2

EVOLUTION

Can we summarise our evolution? Difficult, and I am not an evolutionary mogul, but let's try anyway. I can assure you that even my effort, after reading a lot, could be wildly inaccurate but it may help just to have an idea. If I could put this in a scale I would need all this room. So forget about scales.

								Thousand Years		
Million Years								50,000	10,000	
25	15	5	4	3	2	1				*Today**

Evolutionary Steps

Step 1	Step 2	Step 3
Gathering	Hunting?	Agriculture
1. High food biodiversity	1. Decrease in biodiversity	1. Lower biodiversity
2. High exercise levels	2. Increase in calorie input	2. Higher calorie input
3. Very low calorie input	3. Salt difficult to find	3. Salt easier to find
4. Salt difficult to find		
Ratio:	Ratio:	Ratio:
Effort/calories	Effort/calories	Effort/calories
very high	high	much lower
Effort/proteins	Effort/proteins	Effort/proteins
very high	high	much lower

*In our life today, the ratios effort/calories and effort/proteins are very limited and the availability of salt and sugar unlimited.

6

To get the same daily amount of proteins we get on average today, a gathering-hunting individual only 1,000,000 years ago would need some 18 days. The effort to get his/her food would be some 180 times higher than our effort today.

Chapter 3

THE HUNTING SEASON

The faster and very effective accumulation of proteins and fats is a very important factor when you hunt and the sources of food are either not always available or not predictable. Natural selection has favoured the best accumulators.

If you do not accumulate, you die. If you hunt and kill and rapidly accumulate proteins and fat, you will be better off both in the situations when:

(a) no food is available for a while,
(b) there is a new hunting possibility.

A lot of proteins and fats, acting like high energy fuel, will propel you much more effectively towards the next prey. Therefore, individuals with a very high power of accumulating proteins and fats (and with the capability of quickly using their protein + fat body stores) should be favourably selected in an evolutionary hunting ground.

Also if, for any reason, there is no food available for a while, if you have accumulated your protein + fat stores, you will be a much better survivor.

Fig. 4 Lioness in Tena-Tena. Lions live on a diet based on high levels of fat/proteins which are very effective fuel for rapid and high-power action. Accumulation is a very uncommon problem in these "pure" hunters and "pure" meat-eaters in their natural conditions. To hunt, a lion may travel 10–15 miles a day or swim in a river (I have seen them in Chobe) for 1–2 hours. Sometimes the kill just covers the energy required to hunt.

Chapter 4

ATHERO-PEOPLE

Can we consider the faster/more effective acquisition and storage of fats and proteins in the body (particularly in the arterial wall) as the first step in the evolution of atherosclerosis?

The assumption is that subjects who are able to accumulate faster (who now constitute, due to the favourable evolutionary character, most of our population) are at *much higher risk of developing arteriosclerosis.*

All individual selections must be active before reproduction. Therefore, ideally, the fast accumulator (let's call them athero-people), being more effective hunters and survivors, must have a higher reproductive success. The result is more athero-people. Actually, most of our population is constituted of athero-people.

What happens after reproduction doesn't really matter. So if you get some fat in your coronaries after reproducing, natural evolution is not interested and doesn't take note.

Salt/sugar is very difficult to find, particularly in the primitive, savanna-like environments where most of our evolution took place. Individuals who are able to find, recognise (by effectively tasting them) and are very effective in accumulating and storing them are also favourably selected.

Hypertension may theoretically save you in cases where there is a lack of water and dehydration. Therefore, so many individuals with higher blood pressure had been selected as the result of environmental pressure in situations of prolonged physical effort associated with conditions of low salt and water availability.

If you now have inherited these characters and in your life there is a very low level of physical activity, you are in trouble and may become seriously hypertensive.

Accumulation, Release and Use of Stores

It is possible that there are individuals who are very *fast accumulators* but use very efficiently what they accumulate (therefore, there is no surplus of dangerous material to affect their arteries).

Maybe there are less efficient accumulators but they do not use very efficiently their storage of fats and proteins, and therefore, the material remains there and causes atheromatous changes. Maybe, who knows. We may have these combinations:

	Fat and Protein Accumulation	Release and Utilisation	Relative Risk for Athero Accumulation in Arteries
A	Slow	Fast	Very low
B	Fast	Fast	Medium
C	Slow	Slow	Medium
D	Fast	Slow	High*

Or we can summarise the risk (R) for developing atherosclerosis in four boxes (easier to remember) like this:

		Use and release Slow	Fast
Accumulation	Slow	Medium R	Low R
	Fast	High R*	Medium R

* Athero-people

These characters are genetically controlled. The distribution of the athero-people in a population is possibly a specific characteristic of different populations. Our previous vascular epidemiological studies indicate that different populations (i.e. in Central Italy or in the UK) may have different proportions of athero-people. Also, the weight of environmental factors in different places may be different. This may require different strategies to cope with the most common disease in the most evolved, post-industrial countries.

In summary, the incidence of atherosclerosis is linked to so many factors (genetic-congenital and therefore inborn in our design, and/or environmental factors, i.e diet and lifestyle) that it is still quite difficult to relate the incidence of athero-people to the real prevalence of atherosclerosis in a population.

Chapter 5

ALL SELECTION THAT REALLY MATTERS HAPPENS BEFORE REPRODUCTION

Assumption: After reproduction, there is no natural selection.

Therefore, what happens to us (including atheriosclerosis and cardiovascular disease, which mostly affect us in our adult life after the reproductive period) is not useful from the evolutionary point of view. However, being social animals, humans are also conditioned in their evolution by individual/group interactions even after the reproductive age. For instance, a large number of healthy adults, present in a human society or group, may help the younger (below reproductive age) individuals to reach their reproductive age faster, better and with a lower mortality rate.

This may lead to an increase in their reproductive success, which in turn transmits to the next generation the positive characters.

This is the equivalent of helper bees (or other insects) which are not directly involved in reproduction but help with their activities and behaviours those few individuals in the species specialised for reproduction.

There is no doubt that a healthy long-living father may help his offspring well after their reproductive age. How this could affect evolution and the transmission of positive characters is not really understood, in particular, considering the aspects related to the development of post-reproductive age diseases (such as vascular disease linked to atherosclerosis).

Fig. 5 Hominids in the bush. Most of our recent evolution may have taken place in a temperate, Savanna or bush-like environment. Drawing by Azzurra Ricci from an original diorama at the Natural History Museum in Nairobi (Kenya).

Chapter 6

OUR POPULATION NOW

Our population is still definitely a hunting population (or a population of gathering and hunting individuals) in a definitely non-hunting environment. The hunter nature and design is here, deeply inborn in us, but there is no hunting ground (and related life-conditions) any more.

No Selection Any More?

Possibly, there is no more selection according to our pristine design and evolutionary track. We are slowly going towards a *de-evolution* period during which most of our favourable characteristics, which made us very competitive for hunting-gathering, will be attenuated, diluted and finally lost. Possibly, some other characters will be selected.

The Concept of De-Evolution

The de-evolution process will be (well, it is in action now) much faster than the previous evolutionary period as we manipulate our environment and the selection forces cannot easily touch us. Now, we are even becoming more able to directly manipulate our genetic design.

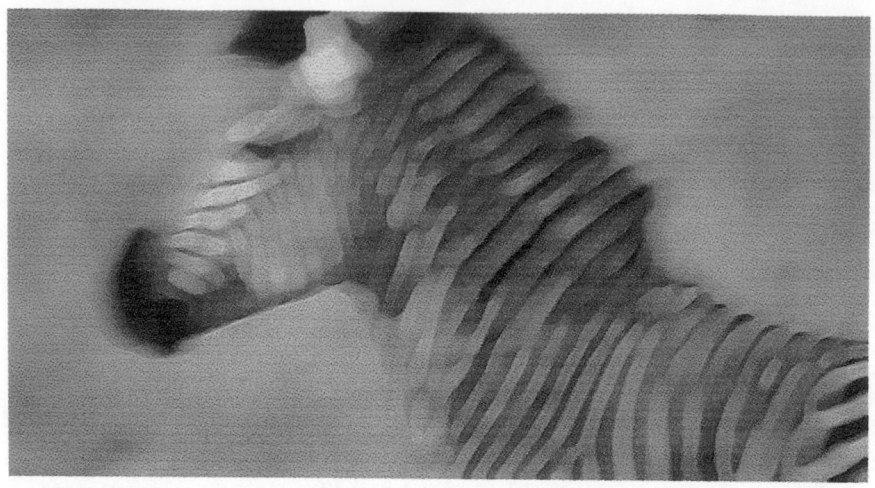

Note: Just to avoid confusion we call "evolution" the process of "natural evolution" as described by Darwin, based on the progressive survival of the "fittest", or possibly the more adapted being to a defined environment. Forced or artificial selection (i.e. by genetic manipulation) is evolution any way and in any sense is "natural" too (as we are part of the evolutionary forces).

Chapter 7

VASCULAR EPIDEMIOLOGY, ITS MEANING AND WHAT POPULATION STUDIES TELL US

Population studies tell us what the most relevant medical problems are in a defined population and how widespread they are. They give us a panoramic view of all medical problems and may suggest preventive solutions.

When some alterations due to a condition affect most people after the reproductive age, this condition cannot be really considered a disease. It is something inborn in the original design of that population. If it is so frequent, it is because it has been selected and perfected as an originally favourable, *evolutionary-produced* character (in the earlier life of the species, in a mostly hunting-gathering life).

San Valentino (Fig. 6) is a great little village in central-east Italy. For many reasons the population of San Valentino represents a typical Italian population. Some 55% of healthy individuals between 40 and 50 years old have some signs of atherosclerosis in their arteries (if you check them with high resolution ultrasound). Between the ages of 51 and 60, some 69% have signs of atherosclerosis and in the following age decade (61–70), most individuals have little or big plaques or thickening of their arteries (85%). In those older than 71, almost all have signs of atherosclerosis. Luckily, only a very limited number of these subjects develop important medical problems.

These observations in the population of San Valentino are comparable (with different proportions of subjects affected by cardiovascular diseases) to the results obtained in the Framingham Study which has been going on

Fig. 6 San Valentino (Central Italy).

in Framingham, a little, peaceful and beautiful town (Fig. 7) just outside Boston (USA) for more than 50 years. The Framingham model to study epidemiology in large populations is a very valid method to evaluate the presence and diffusion of a disease in a community.

Assumption **1**: Arteriosclerosis may be the late evolutionary result of a very favourable selection factor (fast and very effective accumulators of proteins-fats, the athero-people, survive better, hunt-gather more effectively and therefore reproduce more successfully).

Atherosclerosis cannot be a disease if affects so many subjects.

Assumption **2**: Atherosclerosis it is just a status, a pre-disposition to medical problems as so many subjects without any clinical* vascular disease have it and only a very limited number of people really have severe problems.

In one way, it seems that even with atherosclerosis, we survive very well, and only in a very small proportion of subjects, in which the lesions

---

*Means (for non-medical readers) "associated to signs and symptoms (the disease is evident)." Subclinical means that the disease is present but without causing overt signs/symptoms.

Fig. 7 The church at Framingham.

are severe and the disease is dangerous and sometimes deadly. We have adapted ourselves to live well even with atherosclerosis (but this is an erroneous perception as most visible atherosclerosis happens only after the reproductive age and, therefore, it is irrelevant from an evolutionary point of view).

Chapter 8

RIFT VALLEY

November 1946 and After 50 Years

My father left the Rift Valley region in Kenya in November 1946. For five years, he was almost constantly starving, had lost 25 kilos and was nicknamed by his few surviving friends "Baccala" (which means dried codfish) for his appearance. He survived anyway. He was 34 at the time and thousands of his comrades [soldiers firstly and prisoners of war (POWs) after] had died, a few from casualties, but most from the severe, prolonged strain of living in concentration camps. Most survivors had serious problems when they finally left in 1946, including parasites such as amoeba which would have taken years to heal. The many mental strains would stay forever to mark a generation.

He observed and told me three important things:

(1) He had never seen a fat man in the whole of Kenya. All the Masais and other local people he had seen were very skinny.
(2) When there was no food, they (the local people and also those in prisoners' camps) survived eating little quantities of almost anything, including ants and little animals, such as lizards or rats (maybe, he meant rock-iraxes).
(3) Most of the time, in terrible life conditions and in a very difficult environment (however, very similar to our evolutionary playground), he felt in harmony with the world around him.

I was there exactly after 50 years. I took my time, went to a very wild place called Hell's Gate and another called Longonot, which is an incredible

Fig. 8 Panaromic view of the Rift Valley, north of Nairobi.

crater. I just walked alone in the savanna and the bush of these areas looking around.

That was it! Our evolutionary playground again.

After Savuti (in the Okawango delta in Botswana), where I really felt it, I could really feel it again and this time even stronger. The smell of Africa, the African sage, the sounds around. We are all built for that place. I could run and feel it. I was not afraid of animals and they were just keeping away from me.

Harmony, The Hunter Back Home

Try it once if you can. It is simple. It takes a week of your time and some US$4000. Get a plane. Go to Nairobi and get a car if you can. Drive (left side) and go north towards Naiwasha. The Rift Valley and some incredibly spectacular views are on your left but you watch in front of you for trucks as they are quite unpredictable in Kenya. Go to Longonot and Hell's Gate. Walk for a day. Put on a sun-hat and some sun-blocking cream. Try not to be eaten or horned or scratched and remember that you are supposed to be the hunter and not the prey. I strongly suggest staying away from snakes and souvenir sellers. Do not eat funny berries.

I am sure you will feel the same things, experience the hunter's harmony and even write half a paragraph. Ah, do not go in winter. It may be rainy and miserable and you may find out that the weather in London is much better.

Fig. 9 The Masai people at Hell's Gate. The little boy shows the book written by Italian soldiers when they left Kenya in 1946. "Kwa-heri" is farewell in Swahili.

Fig. 10 The Masai village (as a circle of huts) in the crater at Hell's Gate.

Fig. 11 The local doctor shows his "pharmacy".

Lessons from the "doctor":

These populations can teach us a lot. They are the closest relatives to our ancestors, their survival and the preservation of many of their traditions and daily activities is an important priority for us as they may tell us about our original ways of life and our role-model in revolutionary terms.

Chapter 9

A THEORY TO BE DEMONSTRATED

* The theory is that atherosclerosis is not a disease. It is a set of very favourable characters, selected by evolution for a different life.

** We all have some atheromas in our adult age. Only a few of us have medical problems.

*** Years before diffuse atheromas can cause medical problems, we could look for them and positively alter our lives (according to our pristine design) before having medical problems.

**** It is possible that even in the early part of life (before puberty*), we can identify the subjects we can define as **athero-people (fast accumulation and slow release)**. They have characters leading to dangerously high levels of atheromas in their arteries.

---

*i.e. on the basis of genetic tests, or family history or other characteristics.

In these high-risk individuals, we could act very early with education and careful diet/lifestyle control to stop the evolution of atherosclerosis to clinical stages.

Chapter 10

EVOLUTION, NATURAL SELECTION AND ATHEROSCLEROSIS. WE STILL ARE WHAT WE HAVE BEEN SELECTED FOR

No way around it, no bargains. We are very efficient hunters in a non-hunting environment. Our food requirements and food storage characteristics are adapted for a life in the bush or savanna. This ideal life for our design is characterised by:

(1) prolonged, low-energy level physical effort (walking around, gathering);
(2) sudden outburst of force and high-energy requirements (running for hunting or running away from danger);
(3) a very diversified diet, mainly based on very small meals and occasionally (animal kill) on large meals of raw meat;
(4) very little amount of salt and sugar;
(5) a limited number of contacts with other humans.

Compare these points with the equivalent situations in our daily life. The changes in our life have been very fast in an almost zero time, from an evolutionary point of view. We are not really adapted to this new life and we will never be as there is no natural selection any more.

Figure 12 shows the relationship between the amount of physical exercise needed to get enough food to meet our daily energetic requirements at the time of our pristine life and now. Most people use less energy than the amount of daily food provides and therefore store energy like fat. On average in Western industrial populations, this energy excess causes a weight increase between the age of 20 and 60 of some 15–25 kg (some 4 to 6 kg every 10 years).

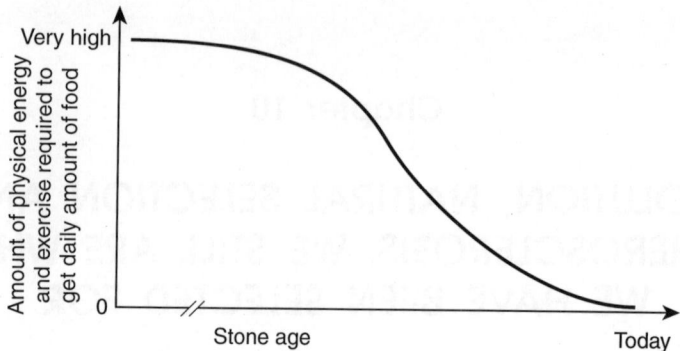

Fig. 12 The amount of energy needed to get the daily amount of food needed to survive at an ideal zero point (beginning of our evolution) during the Stone Age and now (in post-industrial countries).

Chapter 11

NOW WHAT?

We must understand our real nature and tendencies, the structure of our ideal living ground and organise our life according to more natural (to our design) patterns. We can now detect (i.e. with ultrasound) changes in the arteries, years before atheroma causes any disease. The arterial alterations leading to severe problems in the future could be controlled and reversed by changing diet and lifestyle. We can also select (with the same ultrasound* method) from our population those who are very possibly candidates for future cardiovascular problems due to atherosclerosis. We can then act on them.

Therefore we should try to detect as soon as possible athero-people (when they are very young). We may act on them mainly through education, diet and (late stages) with medical treatment, if the risk of imminent cardiovascular disease is too high.

---

*Ultrasound may show a thickening of the inner layers of the arterial walls (i.e. carotids and femoral arteries) or initial, small plaques years before they may cause problems (strokes, infarctions, etc.). These tests are easily available and very cost-effective. One complete ultrasound screening may cost less than US$40–50.

Chapter 12

KNOWING OUR DESTINY

STATEMENT: Once we were hunters. We still deeply are and will be forever (well at least for some million years).

We need to understand this concept and follow our own life programmes according to the original design that has produced us in a very long period of time.

Fig. 13 Bushman graffiti in Savuti (Botswana). Some of the rock paintings in Botswana are 4,000 years old. They all show animals and hunting scenes. An eland, an elephant and a sable antelope are shown here. The life of these people has changed very little in 3,000 years and they live in complete harmony with the surrounding environment.

Fig. 14 The model hunter for humans. A lioness rests after hunting in Tena-Tena. Some animals (i.e. lions) with their group interaction and daily activities may be a role model for our life. The study and evaluation by observation of different species, still living in our pristine environment, may be very useful for reconsidering some behavioural patterns which still survive in our modern society and are an important part of our lives.

Chapter 13

CONCLUSIONS?

These conclusions are really just a beginning.

I hope to stimulate some thinking about the nature of atherosclerosis and its perception as a disease. Can we stop or slow down the evolution of atherosclerosis from its very beginning? Do we really have to wait until individuals become patients? We spend millions to treat and fight atherosclerosis in the final 5–10 years of the disease. We are rarely successful in arresting the disease in this deadly and costly war. Sometimes we temporarily correct a clinical situation (i.e. with a coronary bypass graft) but we cannot really stop the global evolution of atherosclerosis.

But, before the deadly and costly war against atherosclerosis, we have many years of possible diplomacy to alter our lifestyle and go back to the original human design. We are still hunters, forever.

Chapter 14

ATHEROSCLEROSIS AS A RESPONSE TO THE ACCUMULATION OF LIPOPROTEINS IN THE ARTERIAL WALL. THE RESPONSE-TO-RETENTION HYPOTHESIS IN EARLY ATHEROGENESIS

The great affinity of the arterial wall to lipoproteins and its ability to capture lipoproteins is very diffuse among humans (even with variations among different populations). This capability, present in all humans, is one of the basic characteristics of homo.

It is constituted of a set of positive evolutionary characters which has been developed through millions of years of evolution, in constant conditions of scarcity of food containing salt, sugar, lipids and proteins. When these elements became easily and continuously available, the accumulation of fat and lipoproteins in the arterial wall became continuous. The process appears to promote early atherosclerotic changes in the arterial wall and to aggravate their progression.

The beginning of atherosclerosis is still a mystery. There is a theory defined as the *response-to-retention hypothesis* which relates all alterations at the basis of the initial development of atheromas to a process of accumulation in the arterial wall. Many processes have been implicated in the development of early atherogenesis (see Williams & Tabas, *Atherosclerosis, Thrombosis and Vascular Biology* (1995);**15**:551)

(1) endothelial denudation
(2) endothelial injury

(3) endothelial activation
(4) shear-stress-related events
(5) local adherence of platelets
(6) lipoprotein oxidation
(7) lipoprotein aggregation
(8) macrophage chemotaxis and foam cell formation
(9) smooth muscle cell alterations.

It is still not clear which process must be considered as the key event in early atherogenesis (an element which is absolutely required and sufficient as the sole pathological stimulus in an otherwise normal artery) able to provoke the cascade of events leading to athero-formation. As always in biology, there is not a single step but a sequence of actions which work in a chain.

Several reports and experiments, both in humans and animals, now support the idea that subendothelial retention of atherogenic lipoproteins (the accumulation of fats within the arterial wall) is probably the most important initial process in early atherogenesis. The basic idea is that other pathological processes, even important, involved in atherosclerosis are not individually necessary and are not sufficient to begin the formation of an atherosclerotic lesion and plaque.

Most processes detected in atheromas are possibly just a response to the accumulation of fats/lipids in the arterial wall. They may be considered expected biological responses, typical of healthy tissues which, in the presence of retained lipoproteins, react with a series of local reactions linked in a logical chain.

Another competing hypothesis in the formation of early atheroma is the concept of endothelial damage or endothelial denudation (a lesion formed for any reason onto the surface or inside the arterial wall), which is associated with local response to the injury and activation of a reaction of the local cells to the abnormality. This may be defined as the **response-to-injury hypothesis**.

However, there is no clear or definite evidence *in vivo* that endothelial injury is the necessary or sufficient element for formation of the initial arteriosclerotic lesion. In the response-to-injury hypothesis, the idea is that even minimal endothelial damage or desquamation is the key event in the formation of early atherogenesis. However, even in the early stages of their evolution, atheromatous lesions are usually covered by intact endothelium as well as in most of the following stages of progression such as:

- lipoprotein retention
- fatty streak formation
- more advanced atherosclerotic lesions formation.

In humans, only very advanced lesions have no covering endothelial layer. Even in most experimental models, no development of plaque occurs without an intact endothelium. However, the damage could be within the wall structure with a normal endothelium. At this initial lesion, the response may be very individual: some subjects may have no reaction while some may have a very strong reaction with formation of something like scar tissue. It is also possible that the presence of fats within the wall may greatly alter the response.

An improvement of this theory is that many lesions do not cause endothelial damage or denudation but alter the arterial wall causing functional modifications which are the key to atherogenesis (one of the mechanisms is possibly an increased permeability, particularly to atherogenic lipoproteins). This has been defined as the **lipid infiltration hypothesis**.

However, alterations in arterial wall permeability or microscopic damages to endothelial cells are not considered a key requirement for atherogenesis. The normal healthy endothelium transports or leaks out many molecules, including lipoproteins, and the rate of entry of LDL* molecules into a healthy arterial wall exceeds the LDL accumulation rate in the same wall. So, these proteins usually arrive into the wall and, after a while, they are released.

Where are the Lesions?

Pre-lesional (very early) accumulation of atherogenic proteins within the arterial wall is usually concentrated in some arterial sites prone to the development of atheromas/plaques.

---

*low density lipoproteins

The speed and rates of lipoprotein entry into apparently very susceptible parts of the arteries as compared to more atheroma-resistant sites are apparently not too different.

The retention of lipoproteins (and not an increased endothelial or arterial wall permeability) is probably the key factor in the early development of atherosclerosis. Several studies appear to indicate that an increased permeability has a role but this is not essential.

Endothelial permeability may play a role (i.e. in smokers, in subjects with dyslipidaemias and hypertension) only if the infiltrated lipoproteins are retained.

There are other mechanisms occurring later in the evolution of early atherosclerosis.

Lipoprotein retention and aggregation are detected very soon (minutes/hours) after the onset of induced hypercholesterolaemia. The presence of atherogenic lipoproteins regulate endothelial expression of cell-adhesion molecules. The earliest endothelial changes cannot be a cause but are possibly a consequence of the initial retention of lipoproteins within the arterial wall.

Flow irregularities on the arterial wall may also be implicated as they could constitute an early factor causing endothelial trauma. Arterial segments subject to turbulent flow (bifurcations) show a greater predisposition to the development of atheromatous lesions. Many millions of presumed biological alterations have been found in association with flow turbulence (which I will not mention for the sake of your mental sanity).

We have observed that there is a difference in the vasa vasorum network distribution in bifurcations (here, there are less perfusing arteries and less veins). It is also possible that the reduction or absence of lymphatics could impair the removal of lipoproteins when they accumulate in such places. However, this possible explanation is so simple that it will not be understood and accepted by most scientists working in this field.

In vitro studies indicate that there is just a contributory role for shear-stress-induced endothelial alterations. Apparently, shear-stress alterations are not enough and not necessary for determining early atherogenesis. A high concentration of atherogenic lipoproteins is also necessary. Plasma concentration of LDL cholesterol must be higher than 2 mmol/l (80 mg/dl) for atherogenesis even at arterial sites of high shear stress. So, apparently stress-induced endothelial changes may play a limited contributory role in atherogenesis. The most directly related changes at the initial pre-atheroma sites are altered proteoglycan structure and increased lipoprotein retention.

The atherogenic effects of shear stress are dependent on the quantity of lipoproteins within the structure of the arterial wall. It has been postulated that the role of shear stress in early atherogenesis is mediated through the stimulation of intramural synthesis of molecules (proteoglycans) that promote lipoprotein retention. Many stimuli may activate endothelial cells, such as a synergy between shear stress and other mechanisms (i.e. oxidative breakdown of retained lipoproteins).

Other factors are considered potential activators of the endothelium (viruses, homocysteine) but they are not necessary or sufficient for the development of atheromas.

The Second Process at the Basis of Atherogenesis is Considered to be Lipoprotein Oxidation

The oxidation of lipoproteins occurs only after retention within the arterial wall. The process of oxidation is a normal response to lipoprotein trapping. For instance, ApoB* is retained in the human intima before it is oxidised. If oxidised lipoproteins appear in the plasma, they are rapidly removed by the liver and are not deposited in the arteries. Even limited lipoprotein oxidation may be rapidly and strongly atherogenic.

The importance of lipoprotein oxidation has been supported by the claimed effects of antioxidant compounds on atherosclerosis development and progression.

---

*a type of lipoprotein

Retention of Lipoproteins in the Arterial Wall is the Key Event

After rapid injection of LDL-cholesterol (in animals), intramural retention can be detected in minutes. The uptake of lipoproteins in the arterial wall is only part of the problem as an increased wall level may be due to reduced lipoprotein release in sites prone to early lesions in comparison with resistant sites.

Apparently, atherosclerosis cannot easily develop when plasma β-lipoprotein concentrations are low, even in the presence of other major risk factors. The intramural retention of lipoproteins involves the extra-cellular matrix (proteoglycans and other structural elements). Lipolytic enzymes are also involved.

We can summarise the role of all these molecules in atherogenesis as follows:

(1) All these molecules are present in the normal arterial wall and therefore may have a role in the early stages of atheroma.

(2) Retained ApoB is very closely associated with proteoglycans both in early lesions and in more advanced plaques. Arterial proteoglycans (isolated from lesion-prone sites) bind LDL.

(3) LpL* enhances the adherence of LDL to the matrix that is derived from normal endothelial and smooth muscle cells and to normal cell-surface proteoglycans.

(4) A genetic absence of LpL in humans causes hyperlipidaemia without increasing atherosclerosis (possibly, this happens because there is limited generation of small, cholesterol ester-rich particles that are able to enter the arterial wall).

(5) SMase** interacts to cause retention and aggregation of LDL and Lp(a) to arterial cell proteoglycans and matrix.

---

*Well what do you want: they all sound the same. This is another type of lipoprotein.
**This is definitely too much. If you do not understand it, do not bother, most "experts", don't understand either.

Finally, the local factors which may cause and aggravate retention and impair release are not very clear (absence of lymphatics, reduced vasa vasorum network?).

Consequences of Retention

Following retention, LDL undergoes several modifications. Proteoglycan-bound LDL forms aggregates and vesicular structures with increased susceptibility to oxidation. Minimally oxidised LDL induces cells from the endothelium and smooth muscle to express monocyte chemotactic activity. More extensively oxidised LDL attracts monocytes, smooth muscle cells and T-lymphocytes.

Retained LDL is subjected to arterial wall SMase which generates choline phosphate and ceramides (these stimulate mitogenesis and apoptosis). Aggregated or modified LDL is captured by macrophages and smooth muscle cells to form foam cells. Several receptors appear to be involved. Foam cells also stimulate the release of more LpL and other potentially atherogenic factors. Retained, altered lipoprotein stimulates chemotaxis and transformation of smooth muscle cells from contractile to proliferative conditions.

In conclusion, the retention of lipoproteins alone may cause, in a chain reaction, almost all the features of early arteriosclerotic lesions.

Lp(a) constitutes most of the ApoB in human atheromas and may cause many atherogenic effects, including enhanced LDL retention, stimulation of smooth muscle cell proliferation and local inhibition of lysis of microthrombi.

Not All Individuals Have the Same Problems

It is still a mystery why there are so many and wide differences in the development and progression rates among individuals with comparable lipid levels and among different arterial sites. All known risk factors account possibly for just 50% of all coronary events.

The response-to-retention hypothesis predicts that several arterial wall factors are involved in protein retention. The evaluation of these molecules in susceptible arterial sites in comparison with resistant arterial sites is an important research target. Also, it is still a mystery why and how an arterial segment may remain healthy after the entry of lipoproteins. The expulsion of atherogenic lipoproteins from the normal arterial wall is not passive as there are several factors involved in this action.

Arterial retention of atherogenic lipoproteins, a key factor in atherogenesis, is considered as an important, logical target for therapeutic intervention.

Drugs or Others?

Drugs may be developed and used but at a high cost. The wiser strategy is the early individuation by screening of athero-prone people (i.e. by high-resolution ultrasound used to study athero-prone arterial sites, such as carotid and femoral bifurcations). In these people, early primary prevention, aimed mainly at keeping lipoprotein levels as low as possible and at basic lifestyle changes, may be extremely effective at a relatively low cost.

Conclusions

The origin and development of very early atheromas is a complex and multifactorial process. The key problem is the elevated affinity of lipoproteins for the arterial wall while the key pathogenic event is probably lipopotein retention. This event is necessary and sufficient to cause atherosclerotic lesions initiation in an otherwise normal artery. Other potential contributors to the development of early atheromatous lesions are hyperlipidaemia, lipoprotein influx, lipoprotein modification, turbulent blood flow and damages or alterations to the endothelium, smooth muscle cells and matrix. However, these factors individually fail to meet the dual criterion of being necessary and sufficient to produce early atheroma.

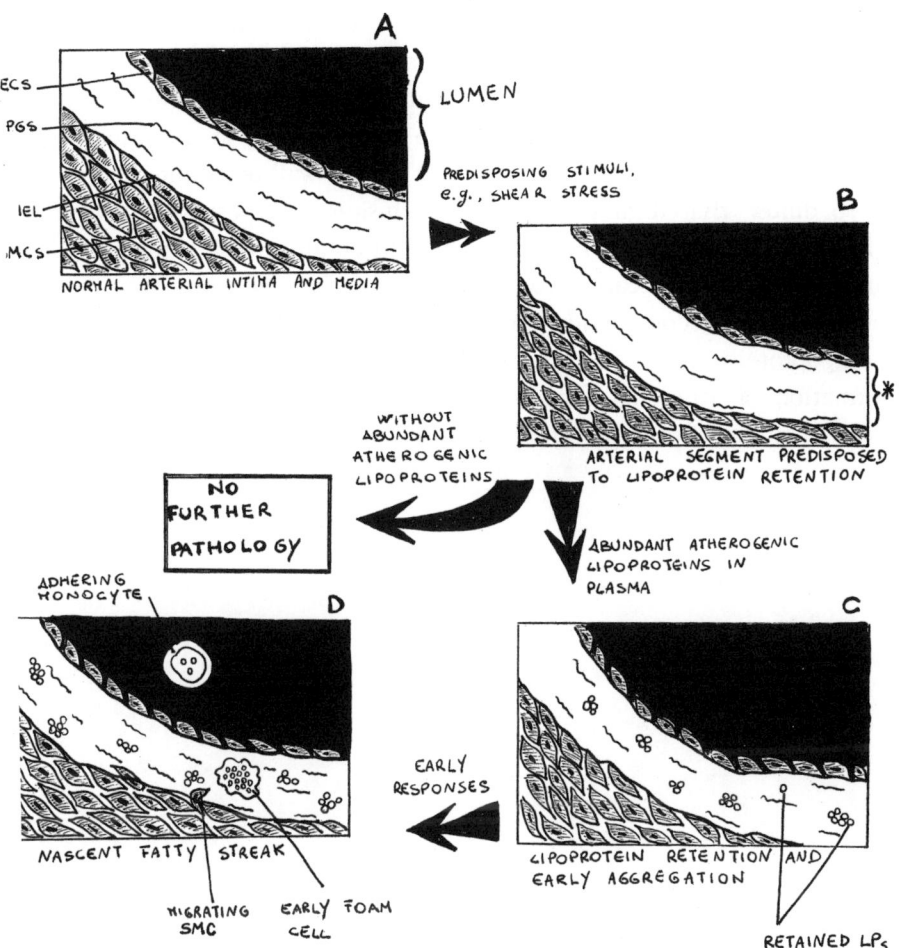

Fig. 15 Chain of events leading to initial plaque formation. Retention into the wall of lipids (fat) starts the chain, eventually leading to plaques causing arterial obstruction.

Therefore, lipoprotein retention may be considered the absolute requirement for lesion development. This is sufficient in most situations to start the complex chain reaction leading to the development of atheroma. All other events and factors involved in atherosclerosis development and progression can be traced back to these early changes in the arterial wall.

In conclusion, the great ability of the arterial wall to capture lipoproteins is so diffuse that it may be basically considered a positive evolutionary characteristic which has been developed in conditions of scarcity of lipids and proteins in our early evolutionary life. When proteins, fats, calories, sugar and salt become easily and continuously available, the accumulation of lipoproteins in the arterial wall is the first determining step in the formation of atheroma. The retention of lipoproteins appears sufficient enough to promote early atherosclerotic changes and their progression.

PART II

EVOLUTION AND MEDICINE

Chapter 15

EVOLUTIONARY MEDICINE

A. Introduction

Evolutionary medicine (also known as Darwinian medicine) concerns the application of the most common evolutionary-based theories to problems related to human health and diseases.

Medicine, as it is often considered and proposed, resembles a mix of information derived from a wide variety of sources, with a few links and without a basic, global linking theory, a rationale or even a generic framework.

Evolutionary theories may be useful, for example, to link easily and in a comprehensible way anatomical patterns and to connect in a logical framework information (i.e. those derived from embryology) with functions of most organs and systems.

The concepts of evolutionary medicine could be used to explain, for example, why the human larynx is structured in a way that it is relatively easy to choke while we could have two separate accesses for feeding and breathing which may avoid choking.

Evolutionary psychology is now used to define and understand the basis of many human behaviours and their links with some mental disorders. Considering evolutionary theories; for example, male/female health differences could be explained in terms of the needed investment males and females have to make to fulfill their respective, interconnected reproductive strategies. The deep divergences in reproductive strategies may now be used for explaining situations associated with sexual dimorphism (i.e. the role of testosterone and other hormonal factors). Behavioural variations between sexes may even be leading to different accident rates in males and females. The origin of deep behavioural differences (even differences in the frequencies of severe violences and homicides) may be associated more to logical evolutionary actions and consequences than to cultural/social conditions which may just be the releasing cause of a deeply rooted behavioural pattern.

For many "experts" (if there are experts in this field), the question of whether natural selection could be definitely relevant to human affairs, such health, diseases and medicine, has already been answered. Evolutionary theories in the evaluation of health problem and evolutionary medicine are definitely very important and relevant to the understanding of diseases.

The real research aims in this field are now to discover in which way, how and in which perspective evolutionary patterns and their consequences are important in the understanding of human physiology, health and medical problems.

In *evolutionary psychology*, the wide adaptive flexibility of the behavioural characteristics of humans has been recognised only recently. Apparently, the reasons for such a large flexibility and adaptability may be linked to the many environmental patterns from which we have evolved. Possibly, we have more instincts and more complex instinctive patterns than most animals competing in the same evolutionary background. This is opposed to the perception that we have fewer instincts when basic human behaviour is considered. Also, in our behaviour, obvious cultural and social patterns often intervene and may cover deeply-rooted instinctive reactions.

The larger the instinct-behaviour menu is available to one animal, the larger is its present adaptability as a result of exposure to a much wider range of environmental pressures during its evolutionary history. Instincts are considered to be specialised cognitive mechanisms that generate, or are the basis, of behaviours. In the past, they have been considered to be more or less fixed, well defined in number and universal in all types of humans. Instincts may also depend on several factors, such as race and geography.

As defined by Bruce Charlton (1997), distinctively human instincts and the resulting behavioural patterns have been produced in response to the millions of years of evolutionary and selection pressures in situations of generally nomadic, foraging hominid society and life in the African or subtropical savanna (or in a similar basic environment).

Evolution Takes Time

For the last 50,000 years, environmental and human social conditions have changed too fast for evolution to follow human development.

Globally, man has evolved faster than the surrounding evolutionary pressures forcing selection and evolution. Humans actually manage to manipulate and alter the environment so much that they finally are becoming possibly the most important, single selection and evolutionary force for all other living organisms on this planet and, luckily or unluckily, for themselves.

B. Original Selection and Evolution, and Discrepancy between the Evolutionary Shape of Humans and the Present Human Figure

Assumption: Many diseases and several conditions leading to health problems may arise from the mismatch between the pristine human conditions, body design, instincts and behaviours and modern life conditions. Most dietary preferences provide a very good example of this discrepancy.

Salt is necessary for life and it was a very scarce resource (something very difficult to find) under ancestral conditions. This is why a strong appetite for *salty foods* has evolved as a positive evolutionary trait (present in all humans, in all races). The capacity of quickly and effectively recognising salt and the strong desire of acquiring it had been an important evolutionary advantage for the primitive humans for millions of years.

The same is true for *sugar*. Ripe fruit is usually sweet and sugary and most often very nutritious while unripe fruits are usually toxic and generally unpalatable. Thus, humans have evolved the capacity of very effectively recognising by taste sugar and sweet substances. On the contrary, unripe fruits are generally sour and unpalatable, and humans have also evolved the taste and sense of smell to rapidly recognise and avoid them.

In modern human life conditions, these strong instinctive appetites lead to the typical negative pattern of pleasurable, mostly obsessive, food purchasing, such as crisps and other salty snacks, chocolate and other sugary things and sweets. Although such preferences do not cause compulsive or uncontrollable eating patterns, they are very important and difficult to control. Also, they are the basis of distinct, negative eating habits in a society where salty and sugary foods are easily available. These negative eating patterns are strongly inborn in all humans, are really universal (unrelated to specific race or culture) and are part of our most pristine human nature.

Therefore, in a totally artificial environment, modern humans need to be educated early in life to control these instincts. They need to learn how to balance their diet, to control their calorie intake and assure for themselves a large food biodiversity because optimal consumption patterns cannot be only controlled on the basis of instinctive behaviour acquired from a bush-like existence.

In the bush-like environment, three factors concurred in forming the diet pattern of the primitive hominid:

(a) the very low calorie intake, irregularly distributed over periods of time, such as days and weeks. The availability of food was greater, mostly constant in tropical and subtropical conditions than in temperate or cold conditions (where long winters could greatly impair food availability);
(b) the time, energy and mental effort needed to gather food or to hunt;
(c) the very large biodiversity of foods available to primitive humans (i.e. the diversity due to seasonal variations).

In contrast, in modern conditions:

(a) the availability of salt, water and calories is high;
(b) the effort (and exercise) to achieve a defined quantity of calories and other elements needed for daily survival is low;
(c) the biodiversity of foods tend to be reduced by several factors (such as packaging, distribution, marketing and cost).

C. Common Objections to Natural-Selection-Based Theories in Medicine

Meanings, scopes and implications of evolutionary theories have expanded since the publication of the theory on natural selection presented in 1859, expecially in its human applications. Nothing could be the same if seen in the light of natural evolution from lovers' dreams to the most terrible war.

Recent advances derived from the natural selection theory include an improved understanding of the evolutionary basis of the major steps in the history and evolution of life on this planet. Also, the "logic" of several common and uncommon diseases and conditions of altered health has been uncovered or at least better defined. It is now apparent that human intelligence and instinct-based behaviour are primarily and mainly designed for enhanced social interaction which is linked to survival. Even the design of our larynx, mainly designed for articulating a large range of sounds, at the risk of choking, is apparently planned mostly for a complex social interaction. This will not happen in other monkeys as they can breath and swallow at the same time but can only produce a limited number of sounds. The articulation and recognition of words (and therefore the concepts behind them) have possibly been a great push in developing some areas of the brain and, by interaction, these areas have stimulated and produced a larger complexity in our vocal system.

Evolutionary theories cannot be indiscriminately applied to all branches of medicine but they may contribute to the general understanding, one may say the philosophy, of a disease or to its manifestations. For instance, diseases such as thrombosis or embolism due to abnormal coagulation may well result from selection pressures for a very effective coagulation in primitive conditions, where cuts and wounds are easily sustained. Individuals with a poor coagulation may have had a very short life in primitive life conditions. The result is a very effective coagulation system, always ready to act, and on the edge of clotting, constantly ready to be activated. In our population, this may frequently lead to lethal, unwanted, not-needed clots (thrombosis).

When we apply evolutionary theories to psychiatry, we find that anxiety, the basis of many disorders, causes several important and frequent categories of problems. Anxiety can be defined as an extreme form of adaptive emotions which are useful for promoting and enhancing attention and to stimulate a complex set of changes in view and anticipation of a stressing event, such as the killing of a prey by primitive hunters. It can be seen how these changes have been slowly and effectively evolved through thousands of years of adaptation to produce a useful state of mind activation which is very helpful (at the right dosage) both in ancestral conditions or in the modern situation of symbolic hunting (i.e. a rugby match, an exam or an interview). But, what if there is no prey, real or symbolic, and the stimulus at the basis of our anxiety is so subtle (or maybe a misleading concept) that it gives us unnecessary worry without a real reason for it?

Maybe (under the light of evolution) that there are two main types of anxiety:

(a) the positive anxiety (i.e. as the one we have in anticipation of a kill or when catching a prey and there is no danger for us)
(b) the negative anxiety (when we may be the prey and there is an imminent danger).

Therefore, the consideration, understanding, diagnosis and treatment of disorders due to anxiety may be changed and seen in a different light.

Applying evolutionary theories to medicine is not simple, as these theories mainly focus and rely, for their demonstration or validation, on complex speculation or on studies of genes and genetic selection.

However, there is more to human nature, health, disease and medicine than genes can explain or justify. Health problems are only an indirect consequence of natural selection. They are often a side effect of the need of the human organism to survive long enough and to compete well enough to attain the highest possible reproductive success relative to their possible rivals.

D. The Evolutionary Design (ED) and the Present Environmental Conditions (EC)

A disease may be due to an external cause breaking the balance between ED and EC. Sometimes, there is an external cause, such as an infection (but we should be already adapted to cope with infections if we have met the agent causing it in our evolutionary past). More often, diseases are due to an altered balance between the original design and the present, modern EC.

The potential and possibilities of evolutionary medicine are in the maximisation and preservation of health based on the knowledge of the origin of a disease and in the full understanding of the meaning of diseases.

In analysing a disease and its "logic", our questions should be (under the light of evolution):

(a) is this an abnormality due to dis-evolutionary situations or behaviours, or a real disease (i.e. an infection not already passed through the evolutionary process)?

(b) can we change the present EC to avoid, attenuate or abolish the disease?

Example 1. Immobility (for instance, forced bed rest due to a disease or bone fracture) causes fibrinogen levels (and other clotting factor levels) to rise and alter their healthy values. This may expose patients to thrombosis. In the primitive life, continuous activity and physical exercise keep the fibrinogen levels to very low values and other clotting factors to healthy levels. Therefore, in patients impaired by forced immobility, we could try to establish a physical exercise plan (or alternatively, we can control coagulation). The first option is more effective, more natural and not costly as the control of clotting factors by drugs may expose patients to side effects and it is usually more expensive.

Example 2. Lack of exercise, altered nutritional habits may lead to weight excess, high cholesterol levels, hypertension and therefore, in time, to atherosclerosis with accumulation of fats and formation of plaques in the arteries. Again, the most natural option is to go back, early in the evolution of the disease, to our pristine design. This means a high level of continuous exercise, low calorie-fat-salt-sugar input and anxiety or stress control. The medical (drug or other) treatment of all these components, basically altered together as a dis-balance between ED and EC, is difficult. Medical control is usually very expensive, very difficult to achieve and control and may cause important side effects. All or most of these problems really arise from the lack of exercise and dietary imbalance, and it is stupid to treat patients without trying to modify their lifestyle.

Standard evolutionary theories cannot always be straightforwardly translated into daily clinical practice and evolutionary medicine requires further work (a lot of work and thinking) to develop and clarify its distinctive viewpoints.

To consider properly and in their best light evolutionary theories and their relation to medicine, we should understand:

(a) evolutionary theories, evolutionary aspects of the human nature and the evolutionary pressures acting on humans in ancestral conditions (which constitute some and possibly more than 99% of human evolution). Our ED has not been modified by the present (lasting not more than 50 years) conditions while our EC has dramatically changed. The ED/EC dis-balance is the key to understanding many medical problems;

(b) the evolutionary basis of health, health problems and diseases in humans;

(c) the contrast between the natural, pristine environment determining human evolution and modern life conditions. These conditions are altered at present in comparison to the original environmental conditions.

Also, the present human behaviours and actions are conditioned by elements which have not been "forecast" or used to force selection by our evolutionary process.

E. Our Life

Modern life conditions, such as a very structured social life (i.e. fixed meals or limited exercise time), the accumulation of food and materials following the beginning of agriculture and structured gathering (i.e. mining, wood collection for fire) and, therefore, the partial or total end of the need for nomadism due to very irregular food availability, involve not more than the last 10,000 years.

This possibly constitutes about 0.1–0.2% or less of our 5–10 (maybe more) million years of evolutionary life. This 0.1–0.2% is quite irrelevant from an evolutionary point of view as such a period is not enough to alter the real human nature.

Us and Today

So here we are. We are supposed to be in the bush mostly walking or strolling, very occasionally running, at least 12 hours a day looking for a little bit of water and something to eat. At the same time, we live trying to stay away from naughty animals with big, nasty looking jaws and claws (also looking for a snack). And here we are in these towns and there is so much food around any way we go.

Reference

Charlton BG. A syllabus for evolutionary medicine. *J Roy Soc Med* (1997), 90, 397.

F. Evolution and Our Story Now. And Cosmically Speaking: Does It Matter?

The Hitch Hiker Guide to the Galaxy

The Hitch Hiker Guide to the Galaxy is a remarkable book for several reasons. The first one is that it is the only existing trilogy which is composed of five books (and possibly more to come) and the second one is that it describes a lot of theories and hypotheses about the origin of the universe, particularly human evolution and origins, considering them more a joke to amuse readers and BBC listeners than anything serious.

On the contrary, most writers on evolution have taken and still take themselves quite seriously, and they may go so far as to explain in a 2000-page book why a butterfly is of one colour instead of another. They do not explain crucial points, such as what is the actual connection between butter, flies and butterflies, which has strangely interested some scientists all their life.

Do not believe them. They do not have a clue. Every six months, there is a new crucial discovery of the first human. We are all derived from that skeleton. At least for a few months. According to the hitch hiker guide, we are all derived from a single race arriving from another remote planet. Their spaceship land-crashed somewhere on earth when humanoids were just off the trees (bored to the bones after talking for million of years of leaves and berries) and had started to have a more interesting life running after animals or being chased by some of them in the bush or savanna. The occupants of the alien spaceship were absolutely similar to our modern shape. They were mostly idiots. The story behind the landing goes back a few years when somebody on the distant planet got really fed up with management, bureaucrats, hairdressers, financial executives, bank and insurance managers and other parasites who were making their lives impossible. We are talking of a distant planet and any relation to equivalent jobs/professions on our planet is absolutely casual and unwanted. These intelligent beings invented the story that the planet was doomed. Why, it is irrelevant (if it was a moon crashing into the sun in a few weeks, or onto the planet itself, or a star-goat going to eat the planet of that solar system, doesn't really matter now).

Therefore, three gigantic spacecrafts were built. One was for the achievers (artists, scientists, leaders, philosophers, thinkers, writers, inventors, intelligent people in general, and so on). One was for the workers (people who really did something like building a house, making bread or a bike or an energy plant, cooking, and so on). The third ark was for the middle people (mostly, time and money wasters, professional idiots, politicians, and so on; see the above paragraph). They were asked to go first so that when the people from the doomed planet would arrive on our planet, they had already essential services and people, such as banks, managers, financial advisers, hairdressers, etc., ready for them. After the middle group (the idiots) left, the rest of the population of that planet had a truly simpler and wonderful life and it turned out that the planet was not doomed at all. It leaked out that it had been a stunt, well organised by some intelligent people fed up of being run all their life by idiots. Nobody on that planet complained and, without financial advisers and banks stealing their money, everybody became enormously rich. Again, please note that we are talking of people from another planet.

The idiots who arrived on primitive earth had as soon as possible 30,000 executive meetings, defined the tree leaf as the legal tender, ran into inflation problems and replaced the local humanoids who were the real result of millions of years of natural selection. The local humanoids soon disappeared and the experimental results of evolution on our planet were completely lost. Basically, we are all derived from the idiots who crash-landed on our planet. The story is not very well supported by experimental data but it makes sense.

(*The Hitch Hiker Guide to the Galaxy*, Douglas Adams, Portland House, New York, 1997)

G. The Evolutionary Ladder

According to human-made philosopy, man is at the top of the evolutionary ladder for several reason, i.e. for inventing things such as cars, refrigerators, traffic jams, X-rays, Coke, the neutron bomb, wars, New York, napalm and Vietnam, weekends, pollution, politicians, talk shows and for many other important achievements.

According to dolphin-made philosophy, they (the dolphins) really are at the top of the evolutionary ladder for exactly the opposite reason (i.e. for not inventing the above quoted inventions and achievements) and therefore living a life without all these nuisances except humans, of course, throwing plastic bags and other rubbish into the sea.

English-born philosophers go so far as to explain that only a race able to invent and understand the complexity and deep meanings of income tax and cricket could be at the top of the ladder of evolution. However, here, evolution went probably so fast that only a minority of those living in England have so far developed enough sense and brain to understand and enjoy cricket and definitely none likes or has developed a taste for income tax.

But there is no doubt that nature and evolution, in their infinite wisdom, are slowly but relentlessly working at it and that in a couple of millions of years, most components of the population will evolve to be able to understand, follow and enjoy them.

There is also one more theory that suggest that both the above theories are rubbish and that there is no evolutionary ladder which may define who or which is at the top of the ladder. Evolution just does what it can, changing a bit of everything on earth now and then, being more or less as whimsical as the New York stock exchange. Also, it is (as the stock exchange) completely obscure, particularly to those who profess themselves as experts. They do not have a clue to what it is happening around them but charge enormous fees for their advice.

The top of the ladder, from a gorilla's point of view, is its world, in total harmony with nature, requiring only a minor part of the resources of the environment around its family. The concept of being at the top of something like the animal kingdom is very human-centric and does not require so much brain. Such a complex social interaction and communication, so many alterations produced on the surrounding environment may be the result of some negative factors. While all animals adapt themselves completely to the surrounding resources and blend themselves with their environment, humans are the only animals who are in constant need to modify the environment to satisfy their needs.

From this point of view, particularly, the need to constantly modify the environment, human are real freaks in the animal world. This evolution is mainly cultural as we have some groups (i.e. the bushmen) still living in complete balance with nature and requiring from the environment only a limited, irrelevant amount of resources. The American Indians were also an example of people living in harmony with the environment.

Therefore, the concept of evolutionary ladders is not really correct and the excessive needs of humans, their impact on the surrounding world may be due to an alteration in our evolutive process in the passage from informal, small social groups to more complex, larger social groups (states, nations) unable to control our behaviour.

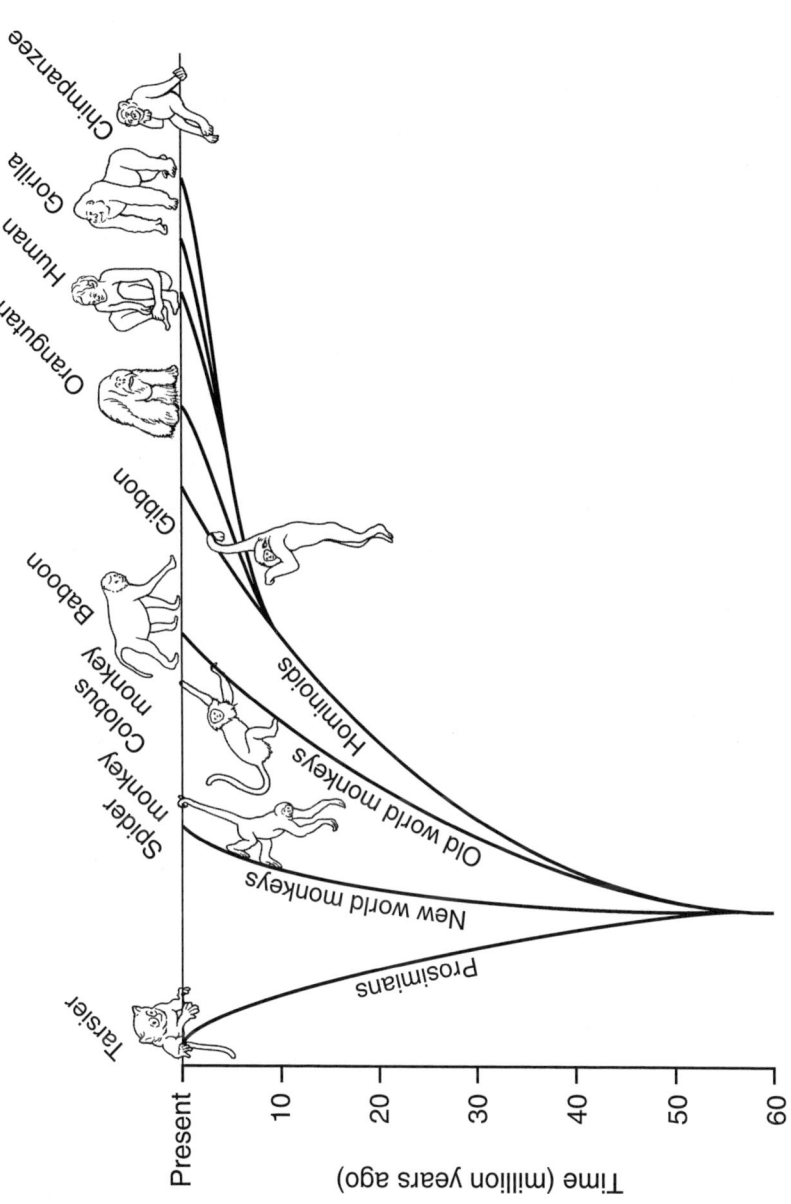

Fig. 16 The family of primates evolved from a primary common ANCESTOR to different species. Many steps in this process are still obscure.

Chapter 16

SAVUTI, BOTSWANA. THE BEGINNING AND THE END OF THE WORLD

Date: one day between some one and three million years ago.

Imagine a large space with sparse vegetation, heat and dust covering the horizon. Elephants (or something similar) move slowly and soundlessly nearby. You also move slowly, on the alert looking for something to eat or little animals that are easy to kill. The water is quite far away now and you have to be sure that you get back there before it is time for the other animals to arrive, particularly the dangerous ones. Yesterday, you had to wait near the little pool close to your, let's call it, home of this week, for more than two hours before you could get a sip of muddy water. A big cat was very close and almost watching it. Better thirsty than being eaten is your simple but effective philosophy. It would be nice to have something to keep some water with you all the time without having to share it with the others.

You walk around carefully listening for any sound which might indicate nasty, moving things also looking around for a meal. Some berries, a flower and some sort of sticky apricot-like things are your shopping or prey today. It is not too much but it could be enough for the day. You are small and skinny and your body does not require much. Today, you are too weak to run after the little antelope you saw before. She is very weak, something is wrong with one of her legs, but still too fast for you and she is still able to keep her distance. Keep an eye on her. She is getting slower and weaker and maybe, if another animal doesn't kill and eat her today, she will be your meal tonight or tomorrow.

Now, you see someone like you but it is not clear from a distance whether it is a relative or someone to keep away from your (more or less your) land.

A. Aggressivity

If you live in an environment like Savuti, you will have many chances of using your aggressivity in small or large measures for two main situations:

— the first is the running-away or escape response
— the second is the hunting response.

The first type of response occurs very often when you hear something moving and you fear it could be something dangerous. It will focus your attention at first (let's call it an anxiety response); then your brain will consider some 3–4 (maybe more) possible responses:

  (i) ignore it, it is not dangerous;
 (ii) run away as fast as you can; it is very dangerous, you cannot cope with it; it is too big, has too many teeth and claws;
(iii) it is dangerous but you can cope with it, you will stay and fight. Maybe, if you are lucky, you could get an extra meal;
(iv) you don't know. Don't move. Hide here, wait and see.

Your day will be a continuous sequence of small or large assorted fears and the consequent brainwork needed to cope with it.

The anxiety response (which is mainly defensive) will be mainly present in your day when conditions of possible, imminent danger for you are present, while the hunting, aggressive response will operate when you think that something is a prey for you and you want to catch it.

During your day, maybe hundreds of times, in minor or major forms, these sensations and states of mind will continuously come and go, making your heart pump faster or slower, continuously finding a way out for emotions.

How different is an environment where the basis of all education and culture is the continuous suppression of these outburst of emotion for which we have been trained and selected over millions of years of Savuti-like existence!

What is the price to pay for controlling the flow of emotional changes within socially acceptable borders? The imminent danger response is still there, anxiety develops but not for something we can really see. Possibly, there is nothing there to trigger our response, it may be a subliminal stimulus we cannot define.

And our wonderful aggressivity, the results of millions of years of hunting, and survival is now just channelled into our tennis hour (once a week) or into symbolic aggressivity such as watching aggressive movies. Outbursts are only socially admitted in rare selected places, situations and times (sports, parties, games).

There is another price to pay in our de-evolution passage from Savuti to New York. The aggressive charge, if not adequately controlled and channeled since early infancy, may cause the most serious problems to our communities, both by its full expression through uncontrolled violence or by its constant and obsessive control or repression which is the basis of most misbehaviours and possibly mental diseases and disorders in our times.

B. The Symbolic Hunting Response

The response to something moving is really innate in most mammals. Two kinds of animals may be considered: predators and non-predators. They are characterised by several, evident body features which may classify them into one of these classes. The most striking is possibly the disposition of the eyes. The predators have frontal, parallel eyes for tridimensional vision, useful to spot and localise the prey, particularly when the prey is running or moving. Non-predators (e.g. rabbits) have eyes at the side of their head, covering almost 360 degrees for vision, mainly developed and used for detecting, while grazing, approaching predators.

Now let's try an experiment. Let's roll a ball close to a cat. It will instantly attract its attention and the cat will start chasing it and playing with it in a symbolic hunt. It will mimick the attack and killing of a prey. Let's do the same with a rabbit. It will run away to a corner of the room trying to escape or, alternatively, to hide and conceal itself. This symbolic hunting response is still deeply rooted in us and may be found in most sports. For example, in football, a group of symbolic hunters in one team runs after a symbolic prey (the ball) in competition with another group of hunters (the competitors in the other team).

This example indicates how deeply rooted are some responses acquired in millions of years of evolution which the present de-evolution (lasting only a thousand years) may conceal and repress most of the time but not completely suppress. Mental control will be achieved only with some mental strain which may eventually show itself under the form of some disorder or disease.

C. Diet and Vascular Disease

Why do we keep eating when we know we are too fat and we do not need food? And why is our willpower to control eating so weak?

Our vascular system is a complex and intricate network which is adapted to carry just the right amount of blood to every part of the body. Most of us today develop deposits of cholesterol and other lipids on and within the walls of the arteries. This often results in strokes and myocardial infarction.

We all know that eating fatty foods and having a non-active life in association with constitutional factors, such as genes predisposing to atherosclerosis, may cause arterial disease leading to strokes and heart attacks. These causes of cardiovascular disease are well known and may be defined as proximate causes. However, recently, we have started to consider what we may define evolutionary causes.

We Have Been Built for a Life with a Little Amount of Food

Humans have been selected in an environment and for a life with constant scarcity of food and salt. Our chronic need of food was a determining factor in keeping us alive in an environment where food was hard to find.

This is why natural selection has not eliminated but favourably selected the individuals with the genes at the basis of the desire for food and fat/sugar/salt. In our present life, where food is readily available, this progressively leads to body accumulation of lipids and cholesterol and their deposition onto our arteries.

Problems due to chronic overnourishment in our age are mainly the result of steady, long-term overeating. For our primitive ancestors, living in the Stone Age or before, it was an adaptive behaviour to pick the sweetest berry or fruit available around. Even this sort of fruit gathering may have cost him some effort and some risk. Imagine the same humans (we are not too different) and what happens when you take people with this very strong adaptation mechanism and put them in a candy shop (our world), full of sweets, cookies, candies, salty snacks and chocolate bars which they can pick up without effort or risk. Almost all will choose these sweets in preference to any modern fruits (such as apple or peach) which are much sweeter (by artificial selection) than any fruit available in the Stone Age.

D. Super-Stimuli

These sweets and chocolate bars may exemplify what the evolutionist defines as *supranormal stimuli*, a misbehaviour described by experts in animal behaviour.

A classic example of over-stimulus can be observed by a simple experiment on geese. If an egg rolls out of a nest, a brooding goose will reach out and roll it back with her chin. Her adaptive behaviour programming is "if an egg or an egglike thing is around the nest, roll it back into the nest". What happens if you put a tennis ball near her nest? She may prefer to roll back the tennis ball. To her it looks more egglike than an egg. So a sophisticated evolutionary response leading to the maximisation of the survival of these animals may result in a totally idiotic and useless response.

Now imagine that there are many equivalent supernormal stimuli in any sensory aspect of our life (for instance, taste). We are continuously confused by our need for sugary and salty things even when we do not really need them for our daily survival. These tastes and desires have slowly and very effectively evolved over millions of years from an almost-starvation situation or severe shortage of food and salt in which we were just able to survive by a narrow nutritional margin. A strong taste for sweet and salt was a crucial survival factor enabling our ancestors to select and eat the fruits/ berries with the highest content in sugar and therefore potential energy, and to find salt which may have been useful to maintain high body-water reserves.

Suggestion: Next time you find yourself reaching for a slice of apple pie instead of a real apple, think of the goose who thinks (well she was actually just following her pre-programming) she should protect and incubate a tennis ball.

E. Conditioning

Most of our dietary problems at the basis of cardiovascular (and other) diseases arise from several pre-conditioned behaviours and by the severe mismatch between the needs and tastes evolved for Savuti-like, Stone Age-like conditions and their effects in our daily life today.

Fats, sugar and salt were in extremely short supply through nearly all of our 4–5-million-year evolutionary history.

The Desire for Today and Tomorrow

In our original world, almost every human, and during most of the time, most of his/her life would almost starve. Therefore, our ancestors may have had important advantages by finding and acquiring larger quantities of these nutritional substances.

The need for them would not stop after eating or drinking a small quantity (what was needed for the day) but it would stay much longer. In this way, it would be possible to accumulate something for an uncertain not-so-lucky tomorrow. Therefore, it was effective and adaptive to want more of them than the daily quantity, to be programmed to desire and to get them in excess for accumulation.

Also, among the human population living in primitive conditions the subjects more capable of accumulating faster and more efficiently fats, sugar and proteins would be much better off in an environment with severe scarcity and irregularity in the availability of nutritional elements.

Today we can afford to eat more fats, sugar and salt than is biologically adaptive, much more than would have been available to our ancestors of only a few thousand years ago. However, it is still very difficult for us to discriminate between the need or desire for the quantity we need now and the quantity we will need tomorrow when the sweets and chips will still be easily available.

With food, stopping at the right point, is not easy for us. Balance and control of the right nutritional quantity and quality could be acquired only by education, which must be used to control our strong innate misbehaviour. It is clear that as a result of evolution, some populations and some individuals are more prone to accumulate and become overweight than others.

Recent studies show that the effort to relieve chronic malnutrition in a population of American Indians (the Pima Indians in Arizona) has caused a diffusion of obesity and of obesity-related diabetes and diseases. Affected individuals have a genetically-based ability to get and store very efficiently food/fat energy. This characteristic is genetically adaptive and constitutes a great evolutionary advantage in an environment characterised by a chronic lack of food and by frequent periods of famine. Individuals with large fat stores survive and hunt/gather better in conditions of chronic food scarcity. The Maoris were able to cross large distances of sea in conditions of relative lack of food (they could fish but the supply of any other kind of food must have been very limited). It is possible to suppose that individuals with a more efficient body accumulation of fats/proteins must have survived better in those long voyages. These subjects, selected over thousands of years of evolution in situations of almost starvation, in this world with easy and continuous access to food may just get fatter and fatter and pathologically over-accumulate fats/proteins and sugar until they reach conditions leading to severe medical problems.

F. Voluntary Food Restrictions

The strong need for food and feeding (particularly for certain kinds of food) in a world offering a lot of opportunities is not easy to correct. Voluntary restrictions may be interpreted by body systems, just as food shortage. For many overweight subjects, it is necessary to completely reset their metabolism to a lower calorie level. This may require some time and food intake control is not easily achieved in a society where the only accepted and traditional exercise for most people is just to have three square meals per day. Also, food restrictions cause increased hunger.

In the Savuti-like environment, the strong stimulus will induce our primitive ancestor to multiply his efforts to find something to eat. In our world, there is no logical relationship between what we feel now, even after having eaten enough, and how much our body needs. Furthermore, many situations of anxiety and stress are erroneously interpreted by our control centres as very similar or equivalent to hunger. The apparent need of food often appears out of the blue when we do not need to eat, with no connection to our nutritional necessity.

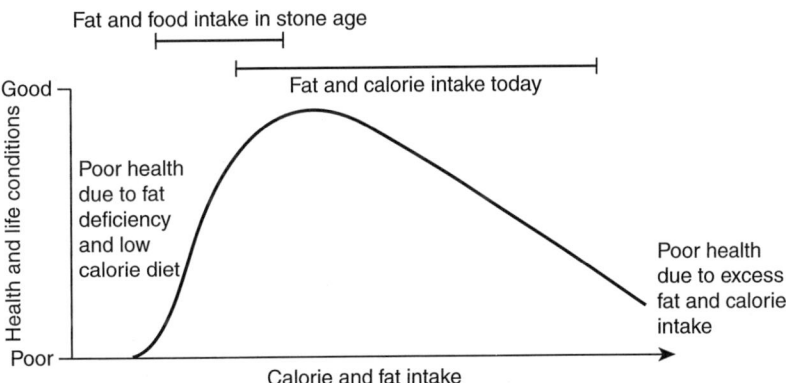

Fig. 17 This key figure shows the relationship between intake of food and nutrients and the benefits produced by these substances. There is a striking contrast between the foraging potential of a Stone Age human and that of a high-salaried diner in a fancy restaurant in New York or London.

It is possible that the need for food and other nutrients could be an evolutionary combination of sensations, including the stress present before getting the food, the psycho-physical effort needed to get it and the positive stress of eating as a reward. If you just get the food from a plate without the preliminaries, you do not really satisfy your control centres which will tell you "Ok, this is enough".

Most preventable diseases in our modern societies results from the devastating effects of a high-fat, high-protein, high-salt/sugar, high-calory diet. Strokes and heart attacks (and possibly cancer too), the most common causes of early death in some social groups, result from arteries and other systems severely altered by atherosclerotic lesions. Health problems result both from decreased food intake and increased food intake.

G. Exercise and Being Lazy

During human evolution, it was an important adaptive mechanism to conserve energy and water by being as inactive and lazy as the circumstances would make it possible. Think of the lions in Savuti. At noon when it is very hot, they just find a place under a tree and wait for hours until it is less hot. The hunting period is generally at night or in the early morning and late afternoon. Apparently, most of the time they are inactive but they are occupied with their most time-consuming activity: saving energy. Most animals do the same in the hottest hours and the strategy (a different one for each species) for saving energy apparently represents one of the most complex evolutionary programmes we have inbuilt.

The *energy-saving behaviours* in primitive conditions are vitally-needed tools as resources (so difficult to achieve in a hostile environment) cannot be wasted. These energy-saving adaptive mechanisms lead to an apparent take-it-easy attitude which forces us to rest as much as possible and to watch sports on television when it could be better for us to do some exercise. This is usually associated and aggravated by our excesses in nutrition.

The average office worker (whose body has been adapted over millions of years of selection to function in a savanna-like environment), would be definitely more healthy if his/her body was used to hunt or harvest food all day with a large utilisation of energy, most often leading to a limited food intake. However, in the case of a possible choice, even the bushman in Savuti would possibly like very much to have an easier life (instead of strolling around all day looking for sips of water and lizards). And he would like a big Mac instead of ants at the end of the day. I am sure he would not give a damn for all these evolutionary theories which involve him being in constant danger of himself suddenly becoming dinner for most big cats.

H. Contacts (Well, Actually No Contacts)

Another striking difference between the life of our ancestors and our life would be the incredible difference between our quantity and diversity of contacts both with other animals and humans.

If you were so unlucky (or lucky, depending on points of view) as to be in Africa some three million years ago, the chance of meeting other individuals (not within your family or tribe) would be remote. A small number of humans would be part of a sparse population in very large spaces. Even considering that you may have walked an average of 15 km every day, you could only have met an average of 1–2 humans every week. However, if you just walked some 5400 km every year (which we can do now, by plane, in an afternoon) you could walk for a total (with an average lifespan of, let's say, 30 years or less) of 164,250 km. Enough distance to run a few times around the planet. However, this was just the case if you walked straight, without rivers, mountains, waterfalls, volcanos, quicksands, and strategic detours around hungry lions or tigers or whatever, including people you did not want to meet at the time for personal reasons and so on.

Apart from Laetoli where the first (or so claimed) human footprints were found perennially (more or less) recorded in volcanic ashes, apparently going quite straight for 10 metres, most humans at that time just strolled or walked slowly around, covering limited distances unless forced by circumstances (i.e. severe scarcity of food and water or moving after a herd).

If you are in the underground in Oxord Circus in central London at peak hours, you may quickly share and blend together all your viruses and bacteria with all other organic life forms around you.

Therefore now, millions of humans get in very close contact often in a limited space or in a close environment. We have not been designed for this situation. It would take thousands of years for one of our Stone Age ancestors to get the same number of contacts. In one hour, now, we could get more contacts than they could have achieved in their lifetime.

In an underground station in Tokyo (Shinjuku), amazed by the incredible number of people around me, I was told that some six million people pass there every day. I disagree. I have been there at peak hours and it looked like a billion. If one out of ten passengers sneezes every day, the equivalent of 600,000 sneezes would diffuse in the closed air system some half a ton of viruses and other biological-rubbish-enriched mucus.

I am also sure that not only the biological contacts affecting our immunology or other system but also the strain of seeing so many humans in the same place at the same time, sharing, what you, by evolutionary conditioning, consider, your vital space is definitely a severe strain for all of us. It definitely costs us a lot in terms of stress and anxiety.

Note: I do not know about the underground in New York as I always use taxis there. Definitely, you get a lower number of contacts by taxi. However, most contacts involve something very weird and (as we are talking about evolution) you get the strong impression that in New York taxis, evolution really went wild. The Museum of Natural History is useful for a glimpse on human evolution. But I am sure you have an interesting, living alternative if you just keep changing cabs.

Conclusions

Contacts, particularly too many unselected and unwanted contacts, are quite unnatural for our evolutionary programmes which generally teach us to keep away from all strangers (look at primitives and children). This is true both considering the immunological responses following so many contacts and stress-related alterations.

We can summarise the problem of contacts.

	Stone Age	Now
Contacts	Very limited	Diffuse, unselected
(Immunology)	Mostly same	Different types
	Family/tribe	of people

Contacts with Other Humans

We have been designed for a Stone Age situation (low level and number of contacts) and not for the present high-contact level condition. How much does this affect our immune system and our psychology (adding a very high level of stress in individuals, particulary those living in congested urban areas)?

I. The Problem of Food Biodiversity

Our ancestor in Savuti would eat very little quantities of an enormous variety of very different kinds of food. Some because he knew that they were food, looked or tasted like food (insects, berries and fruits, snails, animal meat, leaves and so on), some just to try.

At the end of the year, seasonal variations would have produced a large variety of edible (I am not sure they would be edible for us) products. This is so different from our life where mass production of food severely limits the number of different food qualities available in the shops. The quality of food and its biodiversity and freshness are now extremely different from our ancestral conditions.

It is possible that the lack of some accessory foods such as some vitamins, flavonoids or other not well-know accessory nutritional elements (for example, contained in berries) may be at the origin of several common diseases (i.e. varicose veins).

We have definitely been selected over millions of years for a large biodiversity of low-calorie, fresh foods and we end up with almost monodiets based on pre-cooked hamburgers and chips or on what we just find in supermarkets.

There is still an important difference in food biodiversity in small rural villages in Europe in comparison with large towns. In villages, most people have their little vegetable garden, collect, cook and eat snails (sometimes frogs) after a rainstorm. In small villages, it is still common to collect any kind of edible wild herb or root which we will never find in a shop or large market (too expensive to collect and market). In most villages, people will still be strongly linked to seasonal food variations associated with local traditions.

In San Valentino, the village we are screening for cardiovascular diseases (as in most small villages), the biodiversity and seasonal variation of food is much larger than in the nearest town (Pescara) where most people would simply eat what they find in supermarkets. It is possible that the different incidence of cardiovascular disease between places so close and with populations so similar could be associated with accessory food factor supply and with food biodiversity (maybe also with difference in content of vitamins such as E and C).

Most food companies tend to market only those products which are easy to produce, market, pack and sell, which may last a bit longer on the shop shelf and which may maximise profit. Most accessory foods (i.e. berries) have therefore completely disappeared from our diet and town people have completely forgotten about them.

A More Complete Picture

Let's summarise our observations and then compare our life in Savuti and the one in New York or London or any big metropolis now.

For our ancestor in Savuti:

(1) less food
(2) hard to find
(3) less contacts with other people
(4) much more, slow, continuous exercise
(5) much greater psycho-physical effort (a lot of moments of motivated anxiety and outbursts of aggressivity) to obtain the same nutritional value
(6) greater variability and biodiversity of food.

For our metropolitan man:

(1) much more food
(2) easy to find
(3) no effort or physical exercise
(4) millions of contacts per year
(5) monotonous diet composed mostly of the same food all the time
(6) anxiety not linked to real dangers
(7) severely and chronically controlled aggressivity and so on.

J. Immunology and Contacts

A population with a higher level of immunological response to a greater variety of viruses and infective agents may progressively destroy other "immunologically less sophisticated" populations. Charles Darwin (1809–1882) was already aware of this fact and wrote about it in his *Journal of a Voyage Around the World* (the Beagle sailed around the world between 1831 and 1836). This book with all final corrections was published in 1890 and includes a lot of beautiful images (T. Nelson & Sons in London). When he described the natives of Australia, he wrote (verbatim):

Fig. 18 Picture of an Australian hunter from the *Journal of a Voyage Around the World.* After limited contacts with Europeans, most of these populations were wiped away by common European diseases.

Fig. 19 *The Journal of Voyage Around the World* reports the observations collected by Charles Darwin between 1831 and 1836. This edition was published in London in 1890.

Fig. 20 The ship in a phosphorescent sea a little south of the Plata (picture from *The Journal of a Voyage Around the World*).

"The number of aborigines is rapidly decreasing. In my whole ride, with the exception of some boys brought up by Englishmen, I saw only one other party. This decrease, no doubt, must be partly owing to the introduction of spirits, to European diseases (even the milder ones of which, such as the measles* prove very destructive), and to the gradual extinction of wild animals. It is said that number of their children invariably perish in very early infancy from the effect of their wandering life. And as the difficulty of procuring food increases, so must their wandering habits increase. And hence the population, without any apparent deaths from famine is repressed in a manner extremely sudden compared with what happens in civilised countries, where the father, though in adding to his labour he may injure himself, does not destroy his offsprings."

*It is remarkable how the same disease is modified in different climates. On the little island of St. Helena, the introduction of scarlet fever is as dreaded as plague. In some countries, foreigners and natives are as differently affected by certain contagious disorders as if they had been different animals: of which fact some instances have occurred in Chile; and according to Humbolt, in Mexico (Political *Essay*, New Spain, Vol. iv).

In the following page (521) of the book, Darwin expanded this concept quoting other scientists and particularly the work of the Rev. J. Williams, (verbatim):

The Rev. J. Williams, in the interesting work** says, that the first intercourse between natives and Europeans "is invariably attended with the introduction of fever, dysentery, or some other disease which carries off a number of people". Again he affirms: "It is certainly a fact, which cannot be controverted, that most of the diseases which have raged in the islands during my residence there have been introduced by ships. And what renders this fact remarkable is that there might be no appearance of disease among the crew of the ship which conveyed this destructive importation."

**Narrative of Missionary Enterprise, p. 282

Then, he quoted a number of observations by captains of ships, including Captain Beechey, on the effects of epidemics on each island they visited. In a few words, more or less, they arrived and after a while the local population was wiped away.

Apparently, this was also well known to Europeans when conquering America as they probably used common diseases (i.e. flu, cold, measles) as biological weapons to destroy local populations. Is this just a legend? Was it done on purpose? I was unable to find out but it is possible.

Mbarara in Uganda, February 1996

Mbarara is quite uninteresting apart from its good food, nice climate and the fact that you could even get a phone call from Europe in less than one hour.

If, by any chance, you are in Kampala, you could go to Mpigi and from there to Masaka. You need to take your time and some good company as the road is not really fast. From Masaka, go towards Lake Mburo and after the lake you will arrive in Mbarara. At this stage, after a long drive, you need a walk to stretch your legs. Go up the nearest hill and you will find a place called Nkokonjeru Tombs. On your way to the top of the hill, which is not very high, you will find an old mission with a lot of tombs of the young people who arrived there from Europe to help (they thought) the local people and just died after a few weeks of some mysterious disease nobody was able to tell me about.

By the way, if you go up to the top of the hill you will see the tombs of local kings or heroes. From the top of the hill, the view on a nice day is really beautiful with green and yellow hills stretching to the horizon. It is quite an unusual view of Africa and you may imagine you could be anywhere in Europe. Except that the flowers and the vegetation look quite odd (but they are beautiful) and there are no motorways.

Furthermore, in some apparently ideal places (for climate, vegetation, presence of water) in South Africa, cows introduced from Europe died in a short period of time while native animals had survived for years.

Therefore, it is true that the introduction of pathogens from a new species may result in the transmission of a pathogen to which the locals are not accustomed and their immunological system cannot cope with. And it is also possible that the infection or infestation of local species by a pathogen may result in a sort of tolerance by chronic interaction of the pathogen. This may be deadly to species introduced in the area which have not developed a strong immunological response to it.

The resistance to local pathogens/parasites may be an important advantage of certain species to limit the access of other similar competitive species to the same territories and therefore to the some resources. Hence the contacts and the immunological responses acquired in one area may be useless or not very useful in other remote places.

Conclusions

Human species were developed and evolved separately with a very limited number of contacts, creating a sort of immunological nation/island, each with its own characteristics. So many contacts in recent ages, and particularly today, now tend to force these immunological barriers.

K. High Blood Pressure

This could also be considered as a disease or, better, a condition linked to our form of civilisation.

There are several theories considering the origin, problems and complications linked to hypertension, possibly enough to fill a 20,000-page volume. Instead of emphasising the hyper-intake of salt in our diet, we may just consider that blood pressure must be higher mainly to supply the needs of the much larger bodies of humans living in our modern society. We are much heavier and taller than our Stone Age ancestors.

It has been supposed that there is a mechanism that increases blood pressure during adolescent growth spurts. In a savanna-like, prehistoric condition, this mechanism would have made limited, fine adjustments within the range of the small body sizes of primitive humans.

Our hypercaloric, high protein-fat-sugar diet today promotes a very rapid growth and produces much larger body sizes, completely different from the body size which resulted from millions of years of evolution. These present sizes were (and still are) very rare or impossible to achieve with the diet and nutritional conditions available in primordial ages, in a nomadic life. Therefore, the mechanisms regulating blood pressure have been forced to adjust the cardiovascular system and its pressure controls outside the optimum range for that which has been planned by evolution. In larger bodies, pressure may frequently overshoot causing systemic (essential) hypertension and its serious consequences.

Forced Selection

In group of Africans slaves, transported by ship to North America in terrible conditions, only a percentage survived the difficult voyage without enough food and water. Many slaves who died possibly succumbed to malnutrition, diseases and mainly dehydration. It is possible that in chronic dehydration conditions, normal or low blood pressure may have been a cause of death, while in those with higher pressure, dehydration had been better tolerated, allowing survival. Among those who survived the trip, the higher blood pressure could have been transmitted as a dominant character and may have reinforced itself by the artificial, accelerated selection. This contributed to the production of a population with a mean, higher blood pressure and its associated problems. Is it possible that this is one of the reasons why so many black Americans have a higher blood pressure and suffer from its consequences and effects?

Chapter 17

CARDIOVASCULAR DISEASE AND EVOLUTION

The fast accumulation of fats, proteins and sugar in our bodies must have been a very important advantage in the daily activities of our ancestor-prototypes selected over millions of years of evolution (i.e. hunting and food gathering, and maybe running away from the main predators). Also, when seasonal variations in food availability are very large (i.e. low level of food intake in dry or cold seasons, such as winter) the capacity of accumulating long-term stores of fats and proteins is a crucial survival factor in a population living in these conditions. In areas with low temperature and cold, long winter, the faster and more effective accumulation of fats/proteins is also an important protecting factor against cold and starvation.

If you do not have the possibility of accumulating a layer of fats/proteins during the good season, in the following winter, you will soon starve. Also, you will be very cold as the fat layer also has a protective effect from cold. This sort of almost hybernating behaviour must have been acquired by selection in populations living in the North where long, cold winters are the norm. It is possible that if this sort of population is exposed to continuous availability of food, it will accumulate all the time and it will be more prone to atherosclerosis.

In an original, ancestral (1–2 million years ago) subtropical-tropical environment and temperature, the availability of food may have been more or less constant. This may have been an important factor in promoting the evolution of a species such as homo. Local migrations was possibly very limited to following herds of animals in their wanderings. For these populations, the possibility of accumulating fats/proteins was not very important as the availability of food was constant. We may therefore postulate that some Africans races may accumulate (i.e. bushmen) less fats/proteins as a result of the type of environmental pressure.

Food is more or less available during the whole year in a tropical or subtropical environment. African tribes/races living in ancestral conditions still tend to be very skinny. They do not need to accumulate too much of fats or proteins and therefore they are not too efficient as accumulators. On the contrary, for other races (i.e. North European, Eskimo), the capability of constantly and quickly accumulating fats/proteins in summer (when food is available) would be an important advantage. We could define them as pseudo-hibernators. In low temperature and cold winter environments, the accumulation of fats/proteins (in summer) is also an important protective factor against cold.

A. How Do We Use these Observations?

(a) We may postulate that arteriosclerosis is not really a disease.
(b) If a condition affects a very large part of the population, it is not a disease but is a consequence of the evolution which has transmitted these favourable (in original life conditions) characteristics.
(c) Arteriosclerosis is basically the result of the transmission of positive and favourable characters. These have been selected and transmitted through years of evolution.
(d) Different capabilities of fats-proteins accumulations are present in different human races living in different environmental conditions.
(e) The capability of accumulating fats/proteins are possibly linked to environmental stimuli or conditions, promoting or favouring accumulation.
(f) Accumulation rate may be linked to the formation of arterial thickening and plaques and finally to atheroslerosis, causing clinical problems. If we know the characteristics of the population, we may plan different strategies of early detection and prevention of atherosclerosis.

Therefore, arteriosclerosis basically results from a discrepancy between the accumulation power — and the mechanisms at the basis of it, acquired over millions of years of evolution and the large availability of food (associated with the lack of exercise) in our mordern society.

Arteriosclerosis cannot be really considered a disease as it affects most people. Some 85% of people over 55 years of age in the San Valentino cardiovascular screening study have some form (small, limited initial wall thickening or plaques) of arterial alterations. These arterial changes may be observed both in symptomatic and asymptomatic conditions.

An important *limiting factor in food accumulation in our early life* was the fact that crude food could not be preserved for a long period of time and also, it could be dangerous by attracting other animals.

When the simplest food-treating technology became available (first of all fire-cooking), the availability of food and the possibility of accumulating food (by smoking or drying it) increased quite a lot. The limited availability of crude (fresh) food greatly limits the quantity of food intake and fats/proteins accumulation. Just simple cooking of food dramatically increases the possibility of continuous food intake (you can eat cooked food even a few days after a killing) while, particularly in environments of high temperature, food (particularly meat) becomes inedible, if not toxic, after a few hours.

B. Individual and Group Selection

In a modern population, we also have subjects who do not accumulate fats-proteins (or do so very slowly). In a general population, there must be a large percentage of subjects predisposed to accumulation and a smaller number of individuals who do not have this characteristic.

Apparently, evolution works both by individual and group selections. Our evolution has possibly selected some 90% of accumulators (athero-people) and some 10% of subjects who do not accumulate. This is useful to the survival of the species as the differentiation may help in different or changing environmental and evolutionary situations.

The presence of several groups in one species may reflect obvious differentiations and specialisations (i.e. in insects such as ants and bees) and less obvious specialisations such as those visible in humans (useful only in some specific conditions such as shortage of food, diseases resistance to parasites and so on).

C. Evolution — Bits

Biological, Evolutionary-Resulting Accumulation of Food in the Body

The fast accumulation of fats and proteins recently, has been greatly increased by the continuous availability, due to accumulations of fats-proteins outside the body (i.e. refrigerators). The accumulation in the body of biological reserves, which is a positive selection character in conditions of food scarcity, becomes a negative acquisition character when external availability of food is constant and easy.

The continuous desire for food acquisition is useful in conditions where food is difficult to find. However, it becomes an obsession, and therefore a very dangerous risk factor, in conditions of continuous food availability.

Dietary Restriction and Senescence

It has been shown that the large availability of food increases the speed of growth and reproductive success. However, too much food may lead to faster aging. It has been shown (in animals experiments) that it is possible to slow down the aging process by severe diet restriction.

It is also true that important alterations in the diet may produce a series of changes leading to premature aging. The right balance achieved during millions of years of evolution is difficult to find in our post-industrial world today as external conditions tend to be more and more controlled and the availability of food is no longer a strong evolutionary factor for humans in industrialised societies.

Diet-Induced Longevity and Reproductive Success

There is a connection between food availability and accumulation and reproductive success. It appears that those who are able to accumulate better and faster, grow faster, reproduce better or have a better reproductive success. However, as they pass the reproductive age, the advantages given in pre-reproductive age by their characteristics (the results of natual selection) become severe disadvantages (arteriosclerosis, cardiovascular diseases and metabolic problems). Natural selection does not act in post-reproductive life and therefore these disadvantages are irrelevant, from an evolutionary point of view.

In our modern population we have some individuals who accumulate fats and proteins like sponges (we may define them as athero-people). These subjects accumulate fats and proteins very rapidly and effectively and are pre-disposed to arteriosclerosis and to all related cardiovascular problems. They probably constitute more than 60% of the total population.

Preservation of some Negative Characters during Evolution

Some other factors intervene in selection and they are more or less obscure. If some characters are disadvantages (e.g. cystic fibrosis), why are they transmitted and not eliminated? If there is a serious disadvantage, people affected by the problem should die before reaching their reproductive age and therefore the negative character should disappear after some time by negative selection. However, some survive to constitute a part of the population.

Diabetes

The DR3 gene causes early diabetes mellitus but gives advantages *in utero* (fertilised egg adhesion). If a gene decreases the chances of miscarriage, it could be positively selected even if it increases the risk of severe disease in later life.

Childhood-onset diabetes is linked to DR3. If one parent is heterozygous and the other is homozygous for the normal allele, 50% of the babies would be expected to have the DR3 gene. However, the observed rate is 66%. It seems that the presence of the DR3 gene in foetal life greatly decreases the miscarriage rate and, thus, it perpetuates itself, despite causing diabetes later in life.

Many factors (some known and many more unknown to us) are possibly selected as strong positive characteristics for survival during the foetal part of life. They affect the incidence of future diseases as some favourable characteristics during foetal life may be associated with serious defects in pre- and post-reproductive life.

D. How Do We Use These Considerations?

At the moment, we do not really know exactly how to use these considerations, mostly speculations and suggestions with some facts.

Without getting too messianic or founding a new life philosophy, we could suggest some simple steps to avoid, limit and control the negative effects of some evolutionary characteristics on the development and evolution of cardiovascular diseases in particular.

Seven Little Steps:

(1) Try strong educational support to understand the meaning of food needs, consumption, accumulation and eating. Eat when you really need it and avoid ritual eating (most of our eating).

(2) Go back, as much as you can, to natural food biodiversity. Eat according to season and place. Avoid packaged food. After smoking, supermarkets are the strongest risk factor for vascular diseases.

(3) Eat many little meals as irregularly as you can.

(4) Find time and energy for a lot of simple "evolutionary" exercises (no primitive human or bushman will run a marathon or go to a gym to lift weights). No bushman looks like Rambo. Practice a little bit of

chronic exercise as varied as possible. Hunting at the moment is not considered very correct and it is anti-environmental. Anyway there is very little to hunt. If you like hunting and you are a good sport I suggest using just stones and some rare recyclable arrows. Hunt only something you can eat. Please practice possibly in an environment where you could be hunted as well at the same time. Guns are considered anti-sporting and should be avoided in evolutionary hunting grounds and circles. Leave foxes alone. As a good alternative, use a camera to take pictures of animals. In Bwindi (Uganda), we had to walk for hours in the impenetrable forest before we could take some pictures of mountain gorillas. That's a very good exercise.

(5) Get a little bit of stress as soon as you can and as frequently as your life allows without hurting yourself. Lack of stress is a very stressing situation.

(6) Avoid too many useless contacts (taxis are better than the underground, and walking is even better). Stay on your own (or with your closest relative/friends) if you can and as much as you can.

(7) Arteriosclerosis is not really a disease but a condition which may lead to diseases. It is the result of an evolutionary process favouring individuals who are able to accumulate fats and proteins more efficiently and faster in a pristine environment where food was very difficult to find.

Most adult people have arteriosclerotic lesions which may lead to disease at some stage of their life. Therefore, the problem is so widespread that no serious medical solution (e.g. drug) is logically possible at a reasonable cost.

In this field, education is the strongest and most resolutive medicine. It is cheap and simple but there is nothing to gain and therefore nobody is really interested.

Disease produces and wastes money but, anyway, money goes around. It is not wise to say anything about the value of education to prevent cardiovascular disease (in comparison with treatments) as the establishment will consider you a freak to watch with suspicion.

Chapter 18

YOU CAN SAVE YOUR LIFE

As we have seen, many of the positive characteristics present in humans, produced as the result of natural evolution, could become a source of disease in defined conditions not corresponding to the evolutionary ground which has produced that character. For instance, the efficiency of our coagulation system is so high that we can even have a thrombus (a clot) in a vein without any serious trauma or cut. Normally, we should get a clot only if we have bleeding as the first event (a cut, wound or trauma, etc.).

The efficiency of the coagulation system would greatly help a primitive human in a dangerous environment in which he could often and easily get cuts, scratches, frequent minor bruises and wounds, for instance, during hunting or his daily activities. The evolutionary system has produced by natural selection such an efficiency in coagulation because it was better (for millions of years), and still is, to risk an improbable thrombus in comparison with the daily certainty of small cuts and wounds which would cause severe bleeding in case the coagulation was not so effective.

Tissue repair systems and the cicatrisation process have been selected for being present in excess. In the pre-reproductive age, for a hunter at risk of frequent wounds, the ability of quickly closing wounds and to produce efficient scar tissue must have been a very important factor for survival. The ability to produce new tissue needs some mechanisms of careful control to modulate and stop the process at the right moment (when the exact quantity and quality of lost tissue had been replaced and is physiologically effective). This must happen before the new tissue becomes too much, too different or too unusual for the body. The excess of action in the repair processes and the excess in coagulation may be the basis for the development of two of the commonest medical problems: cancer and thrombosis.

Therefore, important evolutionary advantages such as the ability to grow new functional tissue and to produce healthy clots at the right place and at the right moment after the reproductive period could become serious disadvantages if/when the control mechanisims (i.e. due to aging) fail to intervene and control them.

Our species is particularly prone to some sort of serious medical problems and the two most important enemies we now have (as the problem of infections had been largely controlled and abolished) are tumours and cardiovascular disorders. Infections are now considered more an environmental problem but, of course, they may also be a consequence of a weakened body or of an alteration in immunity (patients in critical conditions and with severe immunity alterations have the most serious infections). There are population characteristics, familial and individual differences in susceptibility to cardiovascular diseases and cancers which are still the most serious and the commonest problems. Odd and unusual diseases such as cystic fibrosis and haemophilia run in some families and it is uncommon that somebody affected by this problems does not know it.

However, diabetes which has a strong association with genetics and tend to be frequent in members of the same family, is often forgotten or not considered until it is clinically significant.

Trends in the presence, frequency and severity of the most common diseases, particularly cardiovascular disease which are the leading causes of problems and deaths in our population, should be evaluated and considered, if possible, before they become clinically relevant. Most available medical statistics deal just with the terminal consequences of these problems and it is still difficult to understand the early impact of cardiovascular diseases.

A good plan is to know the role of cardiovascular diseases, in your population, in your community and finally in your family tree.

Just researching on your medical family tree can save your life or give you a warning many years before the medical problem may appear. Creating a simple medical family tree requires that you play the part of a genealogist, detective and researcher at the same time. Talk to your relatives and find, if you can, death certificates, examine medical records (if possible, as often they are not accessible) and get information from other relatives.

It is true that the finding of an illness does not necessarily mean that you are going to get it. You have a lot of genes in common with your first-degree relatives, less with your second-degree relatives (grandparents, aunts and uncles) and fewer with your third-degree relatives (first cousins, great-grandparents).

It is better to have an expert medical opinion (was it the same disease or cancer?) if you find that two first-degree relatives have or had the same cancer (breast, uterine, ovarian and colon cancer should be considered quite similar). Also, get a medical opinion if one or more first-degree relatives under the age of 50 had an illness usually associated with much older people (such as cancers or cardiovascular diseases).

Venous thrombosis may have been present frequently in a family but is often a disease which has been badly reported (a swollen limb, a "phlebitis"). Its consequence (pulmonary embolism) may have been reported as chest pain, chest problems or heart attacks. When a thrombosis has occurred out of the blue in an otherwise healthy subject, it is possible that a coagulation disorder may be involved. This may also be present in other members of the family who may experience the same problem suddenly, without a precise cause, or during periods of higher risk such as immobilisation, illness and surgery. As thrombosis can now be effectively prevented, it is very useful to know, if possible, about the nature and causes of a thrombosis in a relative and whether the problem could have been associated with a hereditary alteration.

The following table suggests some points to investigate concerning the most frequent, possible medical problems in your family and how to deal with them if you find out something.

When you have the information, you need to remember the following points:

(a) Focus firstly on potentially deadly conditions/diseases.
(b) Secondly, focus on diseases which may alter your quality of life (hearing loss, visual problems, etc.).

Find Out	What is the Meaning?
If the relative is deceased when did he/she die, at what age, what was the cause of death?	Date and place of death can produce a death certificate. Death at unusually young age, especially from a particular or rare illness, should be identified. Heart attacks at 45 are very different from the same problem at 75.
How old was the relative when the illness began? cancers	A relative who died at 65 may have been affected by a disease for decades. Inherited diseases and usually occur earlier in life. An early onset disease (i.e. diabetes) may indicate a higher risk for you. Atherosclerosis or thrombosis at an early age are always strange.
What illnesses/health conditions or treatments did they have?	Some people may have lived longer after being treated for serious conditions (cancer, diabetes, etc.) or less dangerous conditions (hearing loss, arthritis) which affected their quality of life.
What was the exact diagnosis? How was he/she treated?	Heart disease may indicate a lot of totally different conditions (hypertension, a cardiac valve problem, infarction, etc.). For any cancer try to understand the first signs and symptoms and the real diagnosis (cancer is vague).
What about risk factors?	Did he/she smoke and drink or eat too much or was there any other risk factor (contact with chemicals) which may be have been the basis of the medical problem?
Is there a story of mental illness?	This problem may run in some families and is often dismissed, kept secret or not mentioned to other relatives.
How many pregnancies were there? Did any child die soon after birth or was born with a birth defect or any congenital problem?	This is another subject which may be forgotten or reported in a distorted way (an abortion could be described as a miscarriage).
What was the person like? What did he look like?	Physical characteristics are very significant. Obesity leads to a series of problems including high cholesterol, hypertension, bone and joint disease. If your relative was fit and exercised regularly and still had heart disease, the problem could have been genetic.
Where did he/she work and live? Was there a dangerous environment which may have caused disease?	The enviroment was often in the past an important cause of morbidity as were social class and education.

(c) Establish (with medical help) a screening plan for the most probable problem which may affect you and your relatives.

(d) Remember cardiovascular diseases are mostly chronic (require a long period of time to develop to clinical stages).

(e) They are easily (in most cases) preventable by education, lifestyle and dietary changes and primary prevention measures.

Note: Generally, cancer prevention is difficult, so you have to focus on early detection and individuation of possible targets. For cardiovascular diseases (particularly diseases linked to atherosclerosis, the most common) there is a good possiblity of early detection. Early detection allows one to reduce or stop progression to clinical stages.

Reference

A. Adato and A. Eccles, Living legacy. Is heredity destiny? Knowing your medical family free can save your life, *Life* (1995) **April**, 60–68.

PART III

LESSONS TO REMEMBER

Chapter 19

A. Junk Food Homo

The great adaptability of humans to different environmental and nutritional situations is an important advantage which allows them to live and thrive in a very large range of geographical and climatic areas. This large adaptability to nutrition is associated with a low selectivity of foods, which is not always recognised by humans as a temporary adaptation mechanism. The fact that humans may easily adapt themselves to live in any environment with almost any sort of food is an important but temporary advantage. However, protracted nutrition with the same kind of food which is not ideal (i.e. almost only based on sugar) leads to severe alterations of the nutritional balance and to diseases (i.e. diabetes) and severe health dangers.

Therefore, this lack of nutritional selectivity in humans, associated with a larger adaptability, also leads to the incorrect use of too much junk food which is the cause of several pathological conditions (i.e. obesity, diabetes, hypercholesterolaemia) and diseases. Most monkeys and baboons do exactly the same (Fig. 21). They adapt their nutrition to the environmental conditions and tend to eat almost any sort of junk (if they have a chance). Monkeys and baboons almost never get cardiovascular disease or a stroke if they live in their normal conditions and eat their "normal", mostly vegetarian diet. A story says that baboons living close to an African safari camp began to eat almost only rubbish from the camp. The rubbish was very rich in fatty, sugary and salty foods (made for the visitors of the camp). One day, one of the monkeys was seen walking in a strange way.

93

Fig. 21 Junk food homo? Most primates may live with almost any kind of food.

She had just had a stroke. Most of the monkeys living close to the camp had arteries with plaques and atherosclerosis.

B. The Cradles

The concept of the "real life", of the pristine environment in which we have evolved during most of our evolutionary life has led to the idea of the *cradles of mankind*, African edens where most of our evolution to the present form may have taken place. Most of the wild places in Africa still resemble our ideal evolutionary ground and are still ideal for indicating to us the measures, limits and cardinal points of our "ideal" life by which and for which we have been produced.

The cultural concept of human cradles, where we can still measure ourselves with the "ideal" evolutionary ground for humans is slowly emerging as an important educational tool to encounter our original way of life.

Fig. 22 Sunset on the Lwanga River close to the Zambesi (Zambia).

Fig. 23 The Lwanga River in Tena-Tena, one of the few remaining spots of 'real' Africa, which are maybe an example of our original evolutionary cradle.

If these areas disappear or are irrevocably altered by human presence or activities, we may lose forever the models on which we should organise our life.

The educational value of these cradles, even if their areas become almost negligible on the planet, should be carefully considered, analysed and shown to children together with their treasures of vegetation and animal life. If we protect and keep these areas as they are, they may still give us, in the future, important indications and suggestions (i.e. concerning our nutrition, new drugs and elements useful for a more natural life).

Some of the "cradles" can still be found in Africa, particularly along the major rivers which have always been important landmarks and hunting places for humans.

C. African Edens

Some of the African Edens are perfectly comparable nowadays to the ideal evolutionary territory where humans spent most of their evolutionary time and may be very interesting for comparing pristine conditions with modern life, if we keep our eyes open. Often, the modern man in the evolutionary ground completely fails to understand the meaning of the world around him and sees the surrounding vegetation and creatures as a human on another planet. His eyes are not used to seeing things with the hunter's eyes.

Most post-industrial people, almost all, would not survive alone in those areas more than 24 hours while the "real humans" (i.e. the bushmen) live in total symbiosis with the environment, finding almost without effort, the resources needed to live, not only to survive. Their life affects nature and its resources only minimally and their passage leaves no visible trace.

I remember that when we crossed the forest in Bwindi to see the gorillas, we left a very large and visible trail of broken branches while the gorillas, three times bigger, advancing faster than us, could move silently and apparently without even touching the leaves and branches around them.

I hope that the divulging concepts relating to an "ideal way of life" would not induce too many people to fly to the remaining cradles, making them like crowded amusement parks and finally changing and destroying

Fig. 24 Rainbow on Victoria Falls one of the original cradles to be preserved for future generations.

their original nature. This type of selective education should be carefully kept separate from package tourism.

I heard somebody speaking of marketing trips to the Sterkfontain caves in South Africa which contain one of the world's most complete and continuous fossil record of the origin of humans, mammals and even dinosaurs! I hope this is not the only way of teaching people about their past and nature.

D. The Lesson of Bwindi, the Impenetrable Forest

Bwindi is the house of our close relative, the mountain gorilla. Looking at these persons (I really do not think we should call them animals), we really have the impression of the complete blending between the gorillas and the surrounding environment. Their powerful shape is in contrast with

Fig. 25 Bwindi, the "forest cradle" reminds us of our past as gatherers when we see the mountain gorilla. The area is just at the corner between Zaire, Rwanda and Uganda.

Fig. 26 The encounter: Our "brother", the silverback mountain gorilla, at Bwindi. Perfect blending with the environment, very specific food range, very defined evolutionary niche and therefore very limited space of survival even if the environment changes very little.

Fig. 27 The cradle: the impenetrable forest at Bwindi (Uganda). Still intact (almost), how long?

Fig. 28 Walking in the impenetrable forest at Bwindi (Uganda) in a food lesson to understand our early "homo" life.

their gentle nature. They look very dangerous but if you go quite close (3–5 metres) you can observe them, hear them "talking", exchanging impressions possibly on different kinds of leaves and on whatever they talk about. I have seen the place where they had slept the night before and the little "cradle" they had prepared (it was a group of eight gorillas) for the two little gorillas of the group.

The gentle strolling, the slow careful, selective and controlled gathering, all made together in a close group are really an example of at least one important part of our nature (the gatherer, as gorillas do not hunt).

E. Our Nature in the Spaceship

If you are in a situation like Bwindi, peaceful and totally different from our artificial urban life, you really gather so much information which will awake many instinctive feelings and sensations still deeply rooted in our design.

I do not say that we have to live like them, it would be difficult and possibly boring. I would very much like to see humans strolling on Mars as soon as possible. However, the deep knowledge which makes the conscience of our real nature, deriving from such encounters, should be always with us, even on spaceships, to help us in our future, post-evolutionary human journey.

F. Chronology

The eras, periods, epochal definitions you find in this book and in books on evolution are schematically represented in the following page (Fig. 29; see also page 55).

All humans apparently derive from prosimians "prototype monkey-like things" present in the Paleocene some 65 million years ago. Most of the present humans possibly originate from the hominoids, and hominids evolved between one and five million years ago in a sub-tropical environment. In the last four million years of evolution, the different branches of hominids evolved, sometimes in continuity and sometimes in parallel. Many human branches eventually disappeared.

Finally, the Erectus emerged as the most successful human prototype. However, most of the other hominids had evolved and had competed in the same or very comparable conditions with overlapping environmental pressures. Why one type emerged as more successful is difficult to understand. It was possibly not something more he had but a larger range of adaptability and therefore a lower selectivity to nutritional and environmental conditions. <u>The most successful humans did not really need an environmental niche. They could find or make one, it did not matter when or where they were. Humans were already slowly, artificially avoiding and escaping the logic and the traps of evolution 1–1.5 million years ago.</u>

Era	Period	Time (millions of years)	Epoch	Cultural stage	Cultural period
Cenozoic	Quaternary		Holocene	Neolithic	Azilian
		0.01	(Upper)	(Upper)	Magdalenian Solutrean Gravettian Aurignacian Chatelperronian
		0.04		(Middle)	↑ Mousterian
		0.15	Pleistocene	Paleolithic	↑ Levalloisian
			(Middle)		Clactonian
		0.5		(Lower)	↑
		1			Acheulian
			(Lower)		
	Tertiary	2 / 5	Pliocene		Oldowan
		25	Miocene	Hominoids, origin of hominids	
		35	Oligocene	Anthropoids, origin of hominoids	
		53	Eocene	Origin of anthropoids?	
		65	Paleocene	Prosimians	

Fig. 29 Timetable of human evolution and culture.

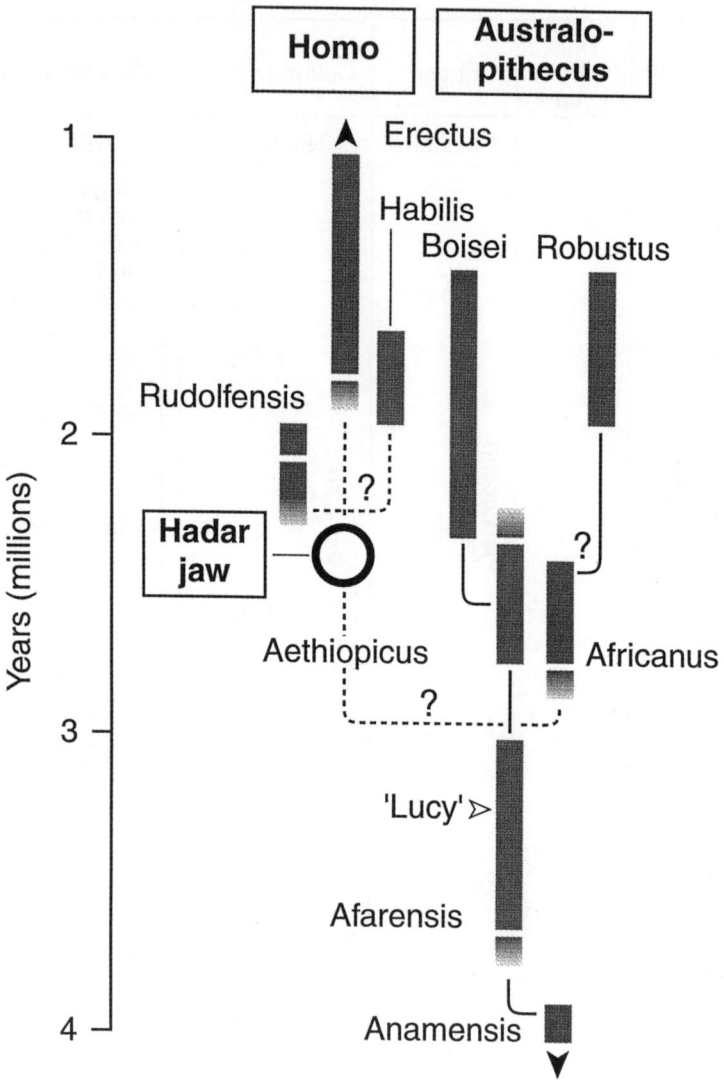

Fig. 30 Evolution of "human" species according to fossil findings and speculations. The picture is still incomplete and it may remain obscure forever. Only the most recent (4 million years) evolutionary life is indicated. The link related to the jaw found in Hadar is also indicated. This jaw may be a common feature of three major lines of evolving humans.

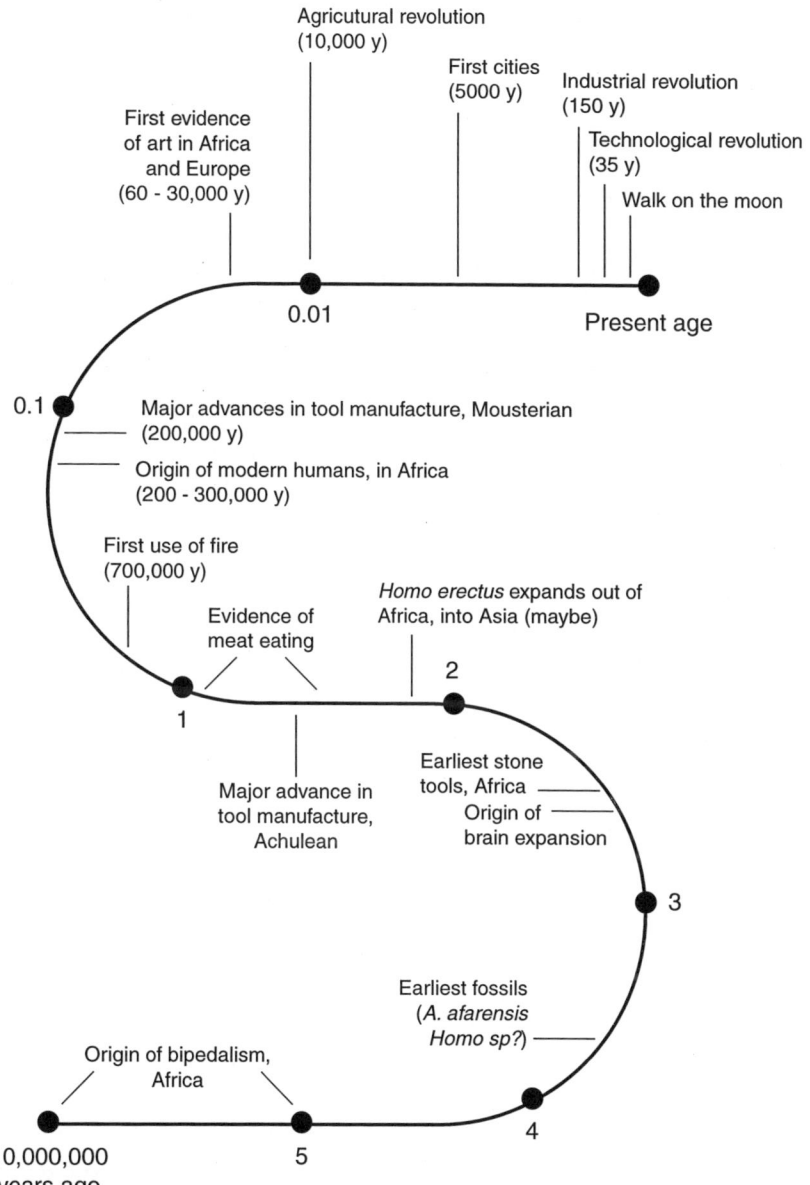

Fig. 31 The evolutionary snake. The last 10,000 years may be defined a period of de-evolution (man modifies environment more than the environment modifies man). No more evolution due to the environmental forces.

G. The Evolutionary Snake

The evolutionary "snake" (Fig. 31) shows the (presumed) steps of human evolution in the last ten million years. This period includes the origin of bipedalism, the manufacturing of tools and the discovery, initial use and then the mastering of fire.

From the first walking in the forest, then leaving the forest for the savanna, to walking on the moon, step after step, the process of evolution first and de-evolution after are in a totally illogical sequence. While other hunters (predators) remained almost unchanged over 10–20 million years, the evolution of humans appears a sort of natural freak, completely different from all other living creatures. The global colonisation of the planet by homo as the predominant living entity is something which is not logically expectable, because in nature, most species live together without predominance in an intricate interactive blending.

Homo predominates since the use of fire and the origin of the modern humans. This happened just an evolutionary second (some 300,000–400,000 years ago, which is just a blink from an evolutionary point of view) after the modern humans take their final shape, and there are major advances in tool manufacturing and social interaction, and possibly language and communication.

H. From the Survival-of-the-Fittest to the Survival-of-the-Fattest

Some 10,000 years ago, the agricultural revolution and animal farming gave humans the ability to pass from occasional gathering to structural and predictable gathering (and the equivalent of structured hunting), increasing both the resources of food and their availability in time. Humans may now partially escape the evolutionary traps linked to the availability of food.

The moment humans find a way of escaping (with tools and fire), partially at first (100,000 years ago), and almost completely most recently, the survival-of-the-fittest race, evolution is definitely outrun.

The creation of the first cities only 10,000 years ago and the creation of structured societies and nations basically ended nomadism and changed the environment in most of our planet forever. Crops were not compatible with wandering animals which were killed or restricted in their migrations almost everywhere. Therefore, the original way of life was changed forever and the de-evolution process irrevocably started.

In conclusion, from the survival-of-the-fittest race in the remote past, the real problem in recent years is now the survival-of-the-fattest.

I. Nomadism, Colonialism and the Evolution of Africa in the Recent Past

The Europeans arriving in Africa, America or Australia found most local populations still living at a nomadic stage. They were not rich, just surviving, almost always on the edge of starvation but they were more or less in tune with the environment they lived in.

Nomadism is the basic life of hunters and gatherers. If you stay in the same place all the time, after a while you either have gathered all you can or killed all the animals around you (and there is nothing else to hunt), or the animals just keep away from you because they are smart. So it becomes more difficult and demanding to hunt. Also, animals do not always stay in the same place. When they have used some of the resources in one place (i.e. grass, leaves, other animals), they move on. Water is not always available in the same places as its availability may vary quite a lot in different places with seasonal changes. Therefore, the ideal gathering-hunting life requires nomadism.

Now, this become difficult or impossible after colonisation of the new worlds. Europeans or European-descendants wanted the land for their crops and animals and did not like the local people or animals wandering freely around. Also, they wanted the local people to produce expensive crops (tea, coffee) with no nutritional values but high gain margins.

Therefore, colonialism completely changed the "ideal life" in the new worlds. The "ideal life" survived only in some remote areas, economically difficult and not too interesting.

Many of the villages in Africa now live in this discrepancy between the European model of towns and the environment which does not allow easy farming (water is difficult to find in many places). Proteins are not easily available as animals keep away from villages. Also, most animals are now protected and local people have no access to their easiest potential source of food. Most money going to national parks from tourists never reaches the local people and even if it does, it is only in minimal amounts.

In most of these villages, the limitation in the variety of crops causes serious nutritional diseases. In Uganda, in beautiful, green fields where any sort of crop could easily grow, most plantations are still tea and coffee, commercially very valuable, and the local people eat very little and very badly.

These populations are not in our world, which they often consider anyway a model, and are not any more in their world (which has been irrevocably changed almost everywhere) where they could live in harmony and with possibly less trouble.

In some places, millions are spent in preserving some rare animals and nothing or very little is done for local people who are a specific, definite and determinant component of the vegetation-animal structure of the place. The prevalence given by pseudo-environmentalist to some television-nice sorts of animals does not really reflect the view of most local people. For example, the protection of elephants has determined an increased number of animals which are real pests for the locals. Most elephants, being really smart, go out of the borders of parks in the evening eating the little vegetable gardens grown with a lot of effort by the local people. It is just a delicacy for them but for locals, it could be a disaster. Giving the local people the possibility of discriminately killing some animals as they always did and the possibility of legally using meat and animal products (i.e. ivory) would possibly stop poaching, improve the life conditions of the local people and keep most areas in a better balance. Also, it could preserve in some areas the heritage of our evolutionary hunting playground where animals and humans live harmoniously together.

These places could be "teaching and learning" areas where we could revive and understand, together with the local hunters, our deepest natural tendencies and needs.

Fig. 32 Little hunters may hunt again. Children play with bow and arrows in Tena-Tena (Zambia). Most of these children had never seen a bow before. Around this village, there is almost nothing to hunt anyway and very little food to gather. It would be nice to see the hunters hunting again with traditional weapons in an "ideal", evolutionary playground.

Fig. 33 The village people changed from hunter-gatheners to farmers (cotton), live a life with a little amount of proteins (available in large quantities, just a few miles away in a national park). I made this bow and the arrow in less than 15 minutes with Natal fig branches. The bow is about 2 metres long and can throw a 1-metre arrow without effort to some 80 metres. It could easily kill an animal at 30–50 metres at no cost.

Fig. 34 This is not my food. The food in the village is palatable and mainly composed of a sort of porridge without or with a very limited amount of proteins. Nutritional problems are evident in most children. Proteins are available just a few miles away. Huge revenues from tourists visiting the local parks never reach the local people. Their priorities, due to an incorrect "pseudo-environmentalist" view of the local problems come well after the priorities of the animals in the parks.

Fig. 35 The elephants could be real pests for local villages. The elephants at Tena-Tena could be a good protein source (there are too many for the area as the number of predators is decreased). However, hunting is still controversial and very difficult for locals.

Fig. 36 Bye-bye evolution. No evolution anymore. The most important evolutionary forces on the planet are now homo and all his activities. ***De-evolution*** is now taking place and while in the previous 65 million years, humans have been shaped by the environment, now the environment is reshaped by homo. The evolutionary design of homo is now complete as it cannot be changed by the environment. The only evolutionary force now active is homo himself, who is able to modify by forced selection or genetic design his structure and function.

Chapter 20

PLACES IN THE BOOK:
THE LAST "CRADLES"

The last African Edens (but there are some areas still unspoilt in other continents) are still in the same locations as original evolutionary grounds for humans.

The places mentioned in the book (Fig. 37) were important both because they were important fossil sites and because (as Savuti or Longonot) they are still environmentally very similar to the pristine conditions in which homo found his evolutionary way to his present shape. Most of the places of interest are close to the breaking line of the Rift Valley in Kenya, the great African lakes from Victoria (north) to Lake Malawi and Victoria Falls in the South. The environment is still very similar to our evolutionary ground in most places in these areas.

Savuti, in the delta of the Okawango, could be a prototype of the ideal, hunting evolutionary ground (savanna and bush) in which most hominids (I imagine) evolved for most of their evolutionary time.

The two separate evolutionary stories of the hominids may be split into consecutive parts:

(a) the first part sees humanoids mainly as gatherers (in the forest, like the gorillas)
(b) the second part sees them in the savanna bush, in open spaces as mainly hunters (very much like lions, hunting in structured groups).

The third part of our evolution is the present de-evolution period in which we have changed our environment and our evolution, possibly all of

Major fossil sites and other places mentioned in the book

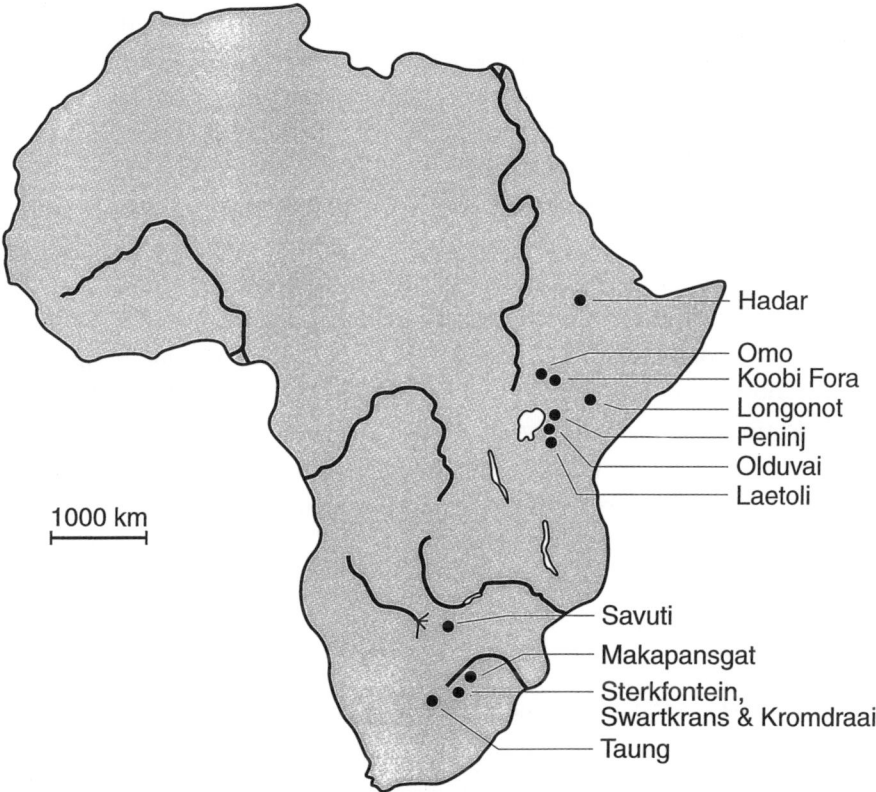

Fig. 37 Major fossil sites and other places mentioned in the book.

evolution, forever. I do not believe that an increase in the population, linked to this de-evolution, will be a real problem, if correctly managed. If the organising social structure is valid and strong, many places could support and feed a much larger population. Holland is the place in Europe with the highest density, still the quality of life is very good in comparison with most parts of the world.

More people in the same place tend to stimulate a strong social effort to accomodate problems and suggest solutions due to continuous interactions, if they forget differences and work on what they have in common. And, well, unless they start killing each other.

The level of the quality of life, at least for us, is definitely much better now than one million years ago when we were a very small component of the animal world. More people mean more connections, more brain power, more solutions, well possibly more problems. At the end of the road, there are no solutions without problems.

The idea of the global village where we all lose our local identities and become, more or less, a uniformly thinking entity, is maybe what we are destined for. By thinking together without silly conflicts, we may be able to put our brain power at work not to exploit differences and fight stupid wars but to challenge the real problems of our world (including animals, plants and all the rest).

I do not know whether we will have a future.

Possibly our future is now. And we need to be aware of our roots, our starting points, our long journey which has shaped us, and that the road ends somewhere. Not too far away.

Imagine Now

Imagine now a land of animals, yellow grass and sparse, tall vegetation, with dry, sandy land and a few waterholes to which animals come to drink.

Imagine yourself as a pristine human now gathering a few things for your daily life and now imagine yourself as the hunter. Find the tracks, stalk the prey and try to kill it. With the simple tools that accompanied you most of your evolutionary time. A bow, some arrows a stone knife.

Then you may walk on the moon and live in artificial universes far away from this world. But you were born there and you will have that nature with you forever.

If you know and understand your origins and your tendencies, you may have a better life and avoid diseases derived from the discrepancy between your pristine design and the present modern life we have created.

If you want to live longer and better, find the homo in yourself.

Once we were hunters... we still are.

INDEX